PEARSON ALWAYS LEARNING

Kirk Trigsted

Guided Notebook for Algebra and Trigonometry

Custom Edition for University of Minnesota Duluth

Taken from:
Guided Notebook for Algebra and Trigonometry, Third Edition
by Kirk Trigsted

Cover Art: Courtesy of Sean Pavone Photo, Jerric Ramos/Shutterstock.

Taken from:

Guided Notebook for Algebra and Trigonometry, Third Edition
by Kirk Trigsted

New York, New York 10013

This special edition published in cooperation with Pearson Education, Inc.

Pearson Education, Inc., 330 Hudson Street, New York, New York 10013
A Pearson Education Company
www.pearsoned.com

Printed in the United States of America

17 2023

000200010272198527

EJ

ISBN 10: 0-135-94954-8
ISBN 13: 978-0-135-94954-2

Algebra and Trigonometry, 3e

Contents

Section R.1 Guided Notebook

Section R.1 Real Numbers

- ☐ Work through Objective 1
- ☐ Work through Objective 2
- ☐ Work through Objective 3
- ☐ Work through Objective 4

Section R.1 Real Numbers

Section R.1 Objective 1: Understanding the Real Number System

What is a **set**?

What is an **element**?

What is used to enclose the elements of a set?

What is used to name a set?

What is a **subset** and how is it written?

What are **Natural Numbers**? Write the first 10 natural numbers.

What are **Whole Numbers**? Write the first 10 whole numbers.

What are **Integers**? Write down 5 positive and 5 negative integers.

What are **Rational Numbers**?

Are all decimals rational numbers? If not, which ones are rational?

Are natural numbers rational? Show why or why not.

Are whole numbers rational? Show why or why not.

Are integers rational numbers? Show why or why not.

What are **Irrational numbers**? Write down 4 examples of irrational numbers.

What are **Real Numbers**?

List the symbols used to denote each of the sets of numbers and which set they represent.

Show the relationship between each of the sets of numbers discussed. Indicate which sets are subsets of another.

Work through Example 1 below. Watch the video to see if your answers are correct.

Classify each number in the following set as a natural number, whole number, integer, rational number, irrational number, and/or real number.

$$\left\{-5,-\frac{1}{3},0,\sqrt{3},1.\overline{4},\sqrt{36},2\pi,11\right\}$$

Section R.1 Objective 2: Writing Sets Using Set-Builder Notation and Interval Notation.

Answer the questions below for the following set $\{x|x<10\}$.

1. Write down how this would be read.

2. Represent this set on a number line.

What is the difference between a solid circle and an open circle?

What is the purpose of the arrow?

Write down the table showing the different types of intervals, their graphs, and how they would be written in set-builder notation.

Work through Example 2 and show work below. Watch the video to check your solution.

Given the set sketched on the number line, a) identify the type of interval, b) write the set using set-builder notation, and c) write the set using interval notation.

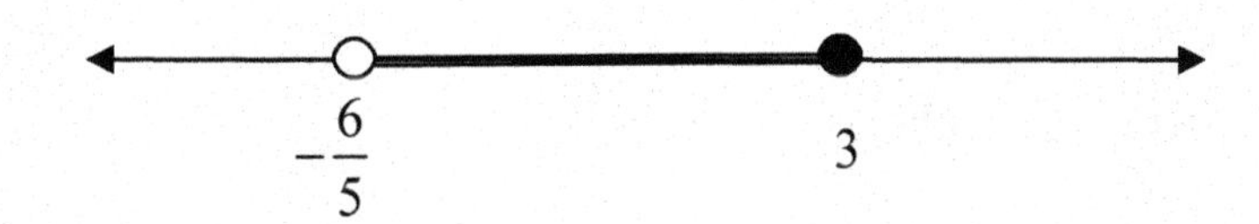

Work Example 3, answering all the questions below, and then watch the video to check your solutions.

a. Answer the following for problem a. Write the set $[-\frac{1}{3},\infty)$ in set-builder notation and graph the set on a number line.

1. What does the set include?

2. How is this written in set-builder notation?

3. Show the graph below.

4. Why is there a solid circle at –1/3?

b. Answer the following for problem b. Write the set $\left\{x \middle| -\frac{7}{2} < x \le \pi\right\}$ in interval notation and graph the set on a number line.

1. How is this set read?

2. Why is there a parenthesis next to –7/2?

3. Show the graph below.

Section R.1 Objective 3: Determining the Intersection and Union of Sets and Intervals.

Fill in the blanks:

When two sets share a _______________ element, we say that the _______________ element is contained in the _____________________ of the two sets.

What is the definition of the an **Intersection** of two sets A and B? Draw a diagram that represents the intersection of two sets.

What is the **empty set** and how is it denoted?

Work through Example 4 showing all work below. Watch the video to check your solution.

Let $A = \left\{-5, 0, \frac{1}{3} 11, 17\right\}, B = \{-6, -5, 4, 17\}$ and $C = \left\{-4, 0, \frac{1}{4}\right\}$.

a. Find A ⋂ B.

b. Find B ⋂ C.

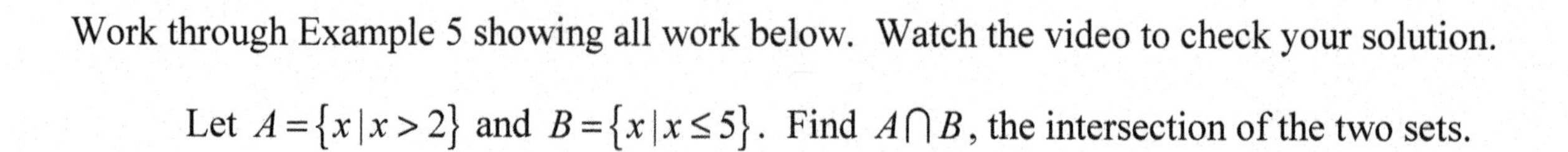

Work through Example 5 showing all work below. Watch the video to check your solution.

Let $A = \{x \mid x > 2\}$ and $B = \{x \mid x \leq 5\}$. Find $A \cap B$, the intersection of the two sets.

What is the definition of the **Union** of two sets A and B? Draw a diagram that represents the union of two sets.

Work through Example 6 showing all work below. Watch the video to check your solution.

Let $A = \{1, 3, 4, 5, 7, 10, 12\}$ and $B = \{2, 4, 6, 8, 10, 12\}$. Find $A \cup B$, the union of the two sets.

Work through Example 7 showing all work below. Watch the video to check your solution.

Let $A = \{x \mid x < -2\}$ and $B = \{x \mid x \geq 5\}$. Find $A \cup B$, the union of the two sets.

Work through Example 8. Check your solutions by watching the video.

a. Answer the following for problem a. $[0, \infty) \cap (-\infty, 5]$

1. Show the graph of the first interval.

2. Show the graph of the second interval.

3. Show the graph of the intersection. Explain the graph.

4. Write the intersection as an interval. Why are brackets around the endpoints?

b. Answer the following for problem b. $((-\infty,-2)\cup(-2,\infty))\cap[-4,\infty)$

1. Draw the graph of $(-\infty,-2)\cup(-2,\infty)$. Why is there an open dot at -2?

2. Draw a sketch of $[-4,\infty)$.

3. Draw a sketch of the intersection. Why is –2 omitted? Why is there an arrow extending to the right?

4. Write the intersection in interval notation. Why is there a parenthesis next to –2?

Section R.1 Objective 4: Understanding Absolute Value

What is the definition of **absolute value**? How is it denoted?

Write down the 5 properties of Absolute Value.

1.

2.

3.

4.

5.

Work through Example 9 showing all work below.

Evaluate the following expressions involving absolute value.

a. $\left|-\sqrt{2}\right|$ b. $\left|\frac{-4}{16}\right|$ c. $\left|2-5\right|$

What is the **Distance between Two Real Numbers on a Number Line**? Show an example.

Work through Example 10 and show all work below.

Find the distance between the numbers –5 and 3 using absolute value.

Section R.2 Guided Notebook

Section R.2 The Order of Operations and Algebraic Expressions

- ☐ Work through Objective 1
- ☐ Work through Objective 2
- ☐ Work through Objective 3

Section R.2 The Order of Operations and Algebraic Expressions

Section R.2 Objective 1: Understanding the Properties of Real Numbers

What is an **algebraic expression**?

What are **variables**?

What is the **Commutative Property of Addition**? Show an example.

What is the **Commutative Property of Multiplication**? Show an example.

What is the **Associative Property of Addition**? Show an example.

What is the **Associative Property of Multiplication**? Show an example.

Work through Example 1 showing all work below. Watch the video to check your work.

a. Use the commutative property of addition to rewrite $11 + y$ as an equivalent expression.

b. Use the commutative property of multiplication to rewrite $7(y + 4)$ as an equivalent expression.

c. Use the associative property of addition to rewrite $(2 + 3x) + 8$ as an equivalent expression.

d. Use the associative property of multiplication to rewrite $8(pq)$ as an equivalent expression.

What is the **Distributive Property**? Show an example.

Work through Example 2 showing all work below. Watch the video to check your work.
Use the distributive property to multiply.

a. $4(x+9)$

b. $(x-4)\cdot 2$

c. $-5(x+y-4)$

Work through Example 3 showing all work below. Watch the video to check your work.
Use the distributive property to rewrite each sum or difference as a product.

a. $xz+yz$

b. $2xw-xy+5xz$

Section R.2 Objective 2: Using Exponential Notation.

Write down the definition of an **exponential expression**?

Write down the CAUTION seen on page R.2-10.

Work through Example 4 showing all work below. Watch the video to check your solution.

a. Rewrite the expression $5 \cdot 5 \cdot 5$ using exponential notation. Then identify the base and the exponent.

b. Rewrite the expression $4x \cdot 4x \cdot 4x \cdot 4x \cdot 4x$ using exponential notation. Then identify the base and the exponent.

c. Identify the base and the exponent of the expression $(-3)^4$, and then evaluate.

d. Identify the base and the exponent of the expression -2^5, and then evaluate.

Section R.2 Objective 3: Using the Order of Operations to Simplify Numeric and Algebraic Expressions.

What is the **Order of Operations**?

Work through Example 5 answering the questions below and showing all work. Then watch the video to check your solutions.

a. Answer the following for problem a. $-2^3+[3-5\cdot(1-3)]$

1. What do you do to everything you're not changing?

2. Show all steps to simplify this expression below.

Answer the following for problem b. $\dfrac{|2-3^3|+5}{5^2-4^2}$

1. Are absolute value bars a grouping symbol?

2. How do you treat the numerator and denominator when there is a large fraction bar?

3. Show all steps to simplify this expression below.

Work through Example 6 answering the questions below and showing all work. Then watch the video to check your solutions.

Evaluate the algebraic expression $-x^3 - 4x$ for $x = -2$.

1. What is the first step to evaluate this expression?

2. Show all steps to simplify this expression below.

What are **like terms**? Give an example of two terms that are like terms and two terms that are not like terms.

Work through the Example 7 video and answer the questions below showing all work.
Simplify each algebraic expression.

a. Answer the following for problem a. $3x^2 - 2 + 5x^2$
    1. What property allows you to rearrange the terms?
    2. Show all steps to simplify this expression below using both techniques in the video.

b. Answer the following for problem b. $-3(4a-7)+5(3-5a)$

1. Show all steps to simplify this expression below.

Section R.3 Guided Notebook

Section R.3 The Laws of Exponents; Radicals

- ☐ Work through Objective 1
- ☐ Work through Objective 2
- ☐ Work through Objective 3
- ☐ Work through Objective 4
- ☐ Work through Objective 5
- ☐ Work through Objective 6

Section R.3 The Laws of Exponents; Radicals

Section R.3 Objective 1: Simplifying Exponential Expressions Involving Integer Exponents

Write down the **properties of positive integer exponents**.

Are they only true for positive integer exponents?

What is the **Zero Exponent Rule**?

What is the **Reciprocal Rule of Exponents**?

What are the 7 **Laws of Exponents**?

Work the problems in Example 1 showing all work below.

Simplify each exponential expression. Write your answers using positive exponents. Assume all variables represent positive real numbers.

a. $-3^{-4} \cdot 3^{4}$

b. $5z^{4} \cdot (-9z^{-5})$

c. $\left(\dfrac{a^{-3}b^{4}c^{-6}}{2a^{5}b^{-4}c}\right)^{-3}$

Section R.3 Objective 2: Evaluating Radicals.

Fill in the blanks below:

A number a is said to the nth root of b if _________.

Every positive real number has ________ nth roots when n is an even integer.

Every positive real number has ________ nth root when n is an odd integer.

We use the symbol $\sqrt[n]{b}$ to mean the ____________________ nth root of b, also called the ________________ nth root of a number b.

The symbol $\sqrt[n]{\ }$ is called a ________________. The integer n is called the ___________. The complete expression $\sqrt[n]{b}$ is called a ________________.

What is the definition of the **Principal *n*th Root**?

Work through Example 2 showing all steps below. Watch the video to check your solutions.

Simplify each radical expression.

a. $\sqrt{144}$

b. $\sqrt[3]{\frac{1}{27}}$

c. $\sqrt[6]{-64}$

Section R.3 Objective 3: Simplifying Expressions of the form $\sqrt[n]{a^n}$

Fill in the blank: If a is a real number and if n is an even positive integer, then $\sqrt[n]{a^n}$ = _____.

Fill in the blank: If a is a real number and if n is an odd positive integer, then $\sqrt[n]{a^n}$ = _____.

Work through Example 3 showing all steps below. Watch the video to check your solutions.

Simplify each radical expression.

a. $\sqrt[7]{z^7}$

b. $\sqrt[4]{x^4}$

c. $\sqrt[6]{(-3)^6}$

Write down the TIP seen after Example 3:

Section R.3 Objective 4: Simplifying Exponential Expressions Involving Rational Exponents

What is the definition of $b^{\frac{1}{n}}$?

Work through Example 4 showing all steps below.
Simplify each expression.

a. $(4)^{1/2}$

b. $(-27)^{1/3}$

c. $\left(\dfrac{1}{64}\right)^{1/6}$

What is the definition of $b^{\frac{m}{n}}$?

Note that the Laws of Exponents apply with rational exponents.

Work through Example 5 answering the questions below and showing all work. Then watch the video to check your solutions.

Simplify each expression using positive exponents. Assume all variables represent positive real numbers.

a. $(9)^{3/2}$

b. $(-32)^{-3/5}$

c. $\dfrac{\left(125x^4y^{-\frac{1}{4}}\right)^{2/3}}{\left(x^2y\right)^{1/3}}$

Section R.3 Objective 5: Simplifying Radical Expressions Using the Product Rule

Write down the **Product Rule for Radicals**.

Write down the CAUTION seen after the Product Rule for Radicals.

Work through Example 6 showing all steps below. Watch the video to check your solutions.

Multiply. Assume all variables represent positive real numbers.

a. $\sqrt{3} \cdot \sqrt{7}$

b. $\sqrt{5} \cdot \sqrt{20}$

c. $\sqrt[5]{9x^2} \cdot \sqrt[5]{14y^3}$

d. $\sqrt[3]{-2} \cdot \sqrt[3]{4}$

Carefully watch the **animation** seen at the top of page R.3-18 to fill in the blanks below:

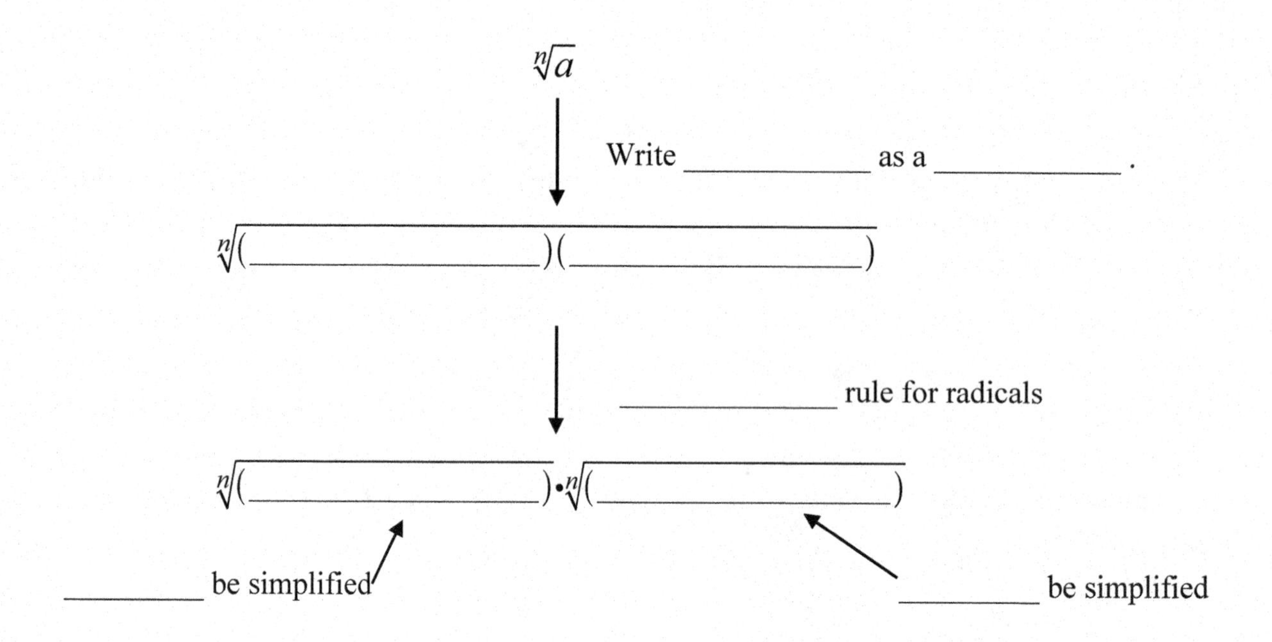

Simplify $\sqrt{12}$ as shown in the animation: $\sqrt{12} =$

Simplify $\sqrt[3]{128a^5}$ as shown in the animation: $\sqrt[3]{128a^5} =$

Write the three steps to

Using the Product Rule to Simply Radical Expressions of the Form $\sqrt[n]{a}$.

Step 1

Step 2

Step 3

Work through Example 7 showing all steps below. Use the product rule to simplify. Assume all variables represent positive real numbers. There is a video solution to part d.

a. $\sqrt{700}$

b. $\sqrt[3]{40}$

c. $\sqrt[4]{x^8 y^5}$

d. $\sqrt{50x^4 y^3}$

Work through Example 8 showing all steps below. Watch the video to check your work.

Multiply and simplify. Assume all variables represent positive real numbers.

a. $3\sqrt{10} \cdot 7\sqrt{2}$

b. $2\sqrt[3]{4} \cdot 5\sqrt[3]{6}$

c. $\sqrt[4]{18x^3} \cdot \sqrt[4]{45x^2}$

Section R.3 Objective 6: Simplifying Radical Expressions Using the Quotient Rule

Write down the **Quotient Rule for Radicals**.

For a radical expression to be **simplified**, it must meet the following three conditions. (Write down the three conditions required for a radical expression to be simplified.)

Condition 1

Condition 2

Condition 3

Work through Example 9 showing all steps below. Watch the video to check your solutions.

Use the quotient rule to simplify. Assume all variables represent positive real numbers.

a. $\sqrt{\dfrac{25}{64}}$

b. $\sqrt[3]{\dfrac{16x^5}{27}}$

c. $\sqrt[4]{\dfrac{7x^4}{625}}$

d. $\sqrt{\dfrac{3x^9}{48x^3}}$

Work through Example 10 showing all steps below. Watch the video to check your solutions.

Use the quotient rule to simplify. Assume all variables represent positive real numbers.

a. $\dfrac{\sqrt{240x^3}}{\sqrt{15x}}$

b. $\dfrac{\sqrt[3]{-500z^2}}{\sqrt[3]{4z^{-1}}}$

c. $\dfrac{\sqrt{150m^9}}{\sqrt{3m}}$

d. $\dfrac{\sqrt{45x^5y^{-3}}}{\sqrt{20xy^{-1}}}$

Section R.4 Guided Notebook

Section R.4 Polynomials

- ☐ Work through Objective 1
- ☐ Work through Objective 2
- ☐ Work through Objective 3
- ☐ Work through Objective 4

Section R.4 Polynomials

Section R.4 Objective 1: Understanding the Definition of a Polynomial

Write down the definition of a **monomial.**

Write down the definition of the **degree of a monomial.**

Write down the definition of the **coefficient of a monomial.**

Complete the table below:

Monomial	Degree	Coefficient
x^5		
$-\sqrt{2}x^2y^7$		
πxyz		

Fill in the blanks below:

A **polynomial** is the ________ or ____________________________ of one or more monomials. A polynomial with two terms is called a ____________________. A polynomial with three terms is called a ______________________. The **degree** of a polynomial is the degree of the ______________________ monomial term.

Answer all the questions for each problem in Example 1. Watch the video to check your solutions.

Determine whether each algebraic expression is a polynomial. If the expression is a polynomial, state the degree.

a. $8x^3y^2 - 2$

b. $5x^5 - 3y^2 - \dfrac{6}{w}$

c. $7p^8m^4 - 5p^5m^3 + 2pm$

Section R.4 Objective 2: Adding and Subtracting Polynomials

What is the procedure to add or subtract polynomials?

Work the problems in Example 2 showing all steps below. Watch the video to check your solutions.

Add or subtract the polynomials as indicated.

a. $\left(6x^4y^2 - 2xy^3\right) + \left(7xy^3 - 8x^4y^2\right)$

b. $\left(6x^4y^2 - 2xy^3\right) - \left(7xy^3 - 8x^4y^2\right)$

Write down the definition of a **Polynomial in One Variable**

Fill in the blank: Polynomials written in descending order are said to be written in ____________________ form.

Work the problems in Example 3 showing all steps below. Watch the video to check your solutions.

Add or subtract as indicated and express your answer in standard form.

a. $\left(-x^5+3x^4-9x^2+7x+9\right)+\left(x^4-5x^5+x^3-3x^2-8\right)$

b. $\left(-x^5+3x^4-9x^2+7x+9\right)-\left(x^4-5x^5+x^3-3x^2-8\right)$

Section R.4 Objective 3: Multiplying Polynomials

How do you find the product of two monomials in one variable?

How do you multiply a monomial by a polynomial with more than one term?

Work through Example 4 showing all work below.

Multiply: $\left(-2x^5\right)\left(3x^3-5x^2+1\right)$

Work through Example 5 showing all work below. Watch the video to check your solution.

Find the product and express your answer in standard form.

$$\left(2x^2-3\right)\left(3x^3-x^2+1\right)$$

The FOIL Method

What does the acronym **FOIL** stand for? Illustrate it.

Work through the **animation** found on page R.4-12 and show how to multiply the following two binomials:

$$(3x+5)(2x+7)$$

Work through Example 6 showing all steps below.

Find the product: $(x+4)(x-4)$

Write down the 5 **special products**.

Work through the problems in Example 7. DO NOT USE FOIL! Use the Special Product formulas. Show all steps. Watch the video to check your solutions.

Find each product using a special product formula.

a. $(2x+3)(2x-3)$

b. $(3x+2y)^2$

c. $(5-3z)^3$

Section R.4 Objective 4: Dividing Polynomials Using Long Division.

Write down each step of the long division process for the problem
$x+8\overline{)2x^2-7x+4}$.

What is the **divisor**?

What is the **dividend**?

Show all steps of the division process below.

What is the **quotient**?

What is the **remainder**?

Watch the video with Example 8 answering the following questions.
Find the quotient and remainder when $3x^4 + x^3 + 7x + 4$ is divided by $x^2 - 1$.

1. How is the divisor rewritten?

2. How is the dividend rewritten?

3. Why do you need to put in the missing coefficients?

4. What must you do in the subtraction step?

5. Complete the problem showing all steps in detail.

Section R.5 Guided Notebook

Section R.5 Operations with Radicals

- ☐ Work through Objective 1
- ☐ Work through Objective 2
- ☐ Work through Objective 3

Section R.5 Operations with Radicals

Section R.5 Objective 1: Adding and Subtracting Radical Expressions

Write down the definition of **like radicals.**

Watch the **animation** seen on page R.5-2 and take notes here:

Work the problems in Example 1 showing all work below. Watch the video to check your solutions.

Add or subtract.

a. $\sqrt{15}+7\sqrt{15}$

b. $10\sqrt{5}-9\sqrt[4]{5}$

c. $\sqrt[3]{\dfrac{4}{27}}+8\sqrt[3]{4}$

Work Example 2 showing all work below. Watch the video to check your solutions.

Add. Assume all variables represent non-negative values.

$$12\sqrt[3]{7x^2}+4\sqrt[3]{7x^2}$$

Work Example 3 showing all work below.

Add or subtract.

a. $\sqrt{54}+6\sqrt{72}-3\sqrt{24}$

b. $\sqrt[3]{24}-\sqrt[3]{192}+4\sqrt[3]{250}$ (Watch the video to check your solution to part b.)

Work Example 4 showing all work below.

Add or subtract. Assume all variables represent non-negative values.

a. $\sqrt[3]{27m^5n^4}+2mn\sqrt[3]{m^2n}-m\sqrt[3]{m^2n^4}$

b. $2a\sqrt{16ab^3}+4\sqrt{9a^2b}-5\sqrt{4a^3b^3}$ (Watch the video to check your solution to part b.)

Work Example 5 showing all work below. Watch the video to check your solutions.

Add or subtract. Assume all variables represent non-negative values.

a. $\frac{\sqrt{45}}{6x} - \frac{4\sqrt{20}}{5x}$

b. $\frac{\sqrt[4]{a^5}}{3} + \frac{a\sqrt[4]{a}}{12}$

c. $\frac{3x^3\sqrt{24x^3y^3}}{2x\sqrt{3x^2y}} - \frac{x^2\sqrt{10xy^4}}{\sqrt{5y^2}}$

Section R.5 Objective 2: Multiply Radical Expressions

Work through Example 6. Multiply. Assume all variables represent non-negative values.

a. $5\sqrt{2x}\left(3\sqrt{2x}-\sqrt{3}\right)$

b. $\sqrt[3]{2n^2}\left(\sqrt[3]{4n}-\sqrt[3]{5n}\right)$ (There is a video solution to part b.)

Work through Example 7. Multiply. Assume all variables represent non-negative values.

a. $\left(7\sqrt{2}-2\sqrt{3}\right)\left(\sqrt{2}-5\right)$

b. $\left(\sqrt{m}-4\right)\left(3\sqrt{m}+7\right)$ (There is a video solution to part b.)

Work through Example 8. Multiply using special product formulas. Assume all variables represent non-negative values.

a. $\left(\sqrt{y}+3\right)\left(\sqrt{y}-3\right)$

b. $\left(3\sqrt{x}-2\right)^2$ (There is a video solution to part b.)

Section R.5 Objective 3: Rationalize Denominators of Radical Expressions

What is **rationalizing the denominator**?

Write down how to **Rationalize a Denominator with One Term.**

Work through Example 9 showing all work below. Watch the video to check your solutions.

Rationalize the denominator. Assume all variables represent non-negative values.

a. $\dfrac{\sqrt{7}}{\sqrt{5}}$

b. $\sqrt{\dfrac{3}{10x}}$

Work through Example 10 showing all work below. Rationalize the denominator. Assume all variables represent non-negative values.

a. $\sqrt[3]{\dfrac{11}{25x}}$

b. $\dfrac{\sqrt[4]{7x}}{\sqrt[4]{27y^2}}$ (There is a video solution to part b.)

Work through Example 11 showing all work below. Watch the video to check your solutions.

Simplify each expression first and then rationalize the denominator. Assume all variables represent non-negative values.

a. $\sqrt{\dfrac{3x}{50}}$

b. $\dfrac{\sqrt{18x}}{\sqrt{27xy}}$

c. $\sqrt[3]{\dfrac{-4x^5}{16y^5}}$

Write down how to **Rationalize a Denominator with Two Terms.**

Work through Example 12 showing all work below.

Rationalize the denominator. Assume all variables represent non-negative values.

a. $\dfrac{2}{\sqrt{3}+5}$

b. $\dfrac{7}{3\sqrt{x}-4}$

c. $\dfrac{\sqrt{y}-3}{\sqrt{y}+2}$

Section R.6 Guided Notebook

Section R.6 Factoring Polynomials

- ☐ Work through Objective 1
- ☐ Work through Objective 2
- ☐ Work through Objective 3
- ☐ Work through Objective 4
- ☐ Work through Objective 5

Section R.6 Factoring Polynomials

Factoring is the reverse of ________________________________.

Section R.6 Objective 1: Factoring Out a Greatest Common Factor

Watch the video with Objective 1 and answer the following questions.

What is the distributive property?

Work the first example in the video below. (Factor: $ab + ac$)

In the second example show how he rewrites each term as a factor of primes.
(Example: $4x^2 - 6x^3 + 2x$)

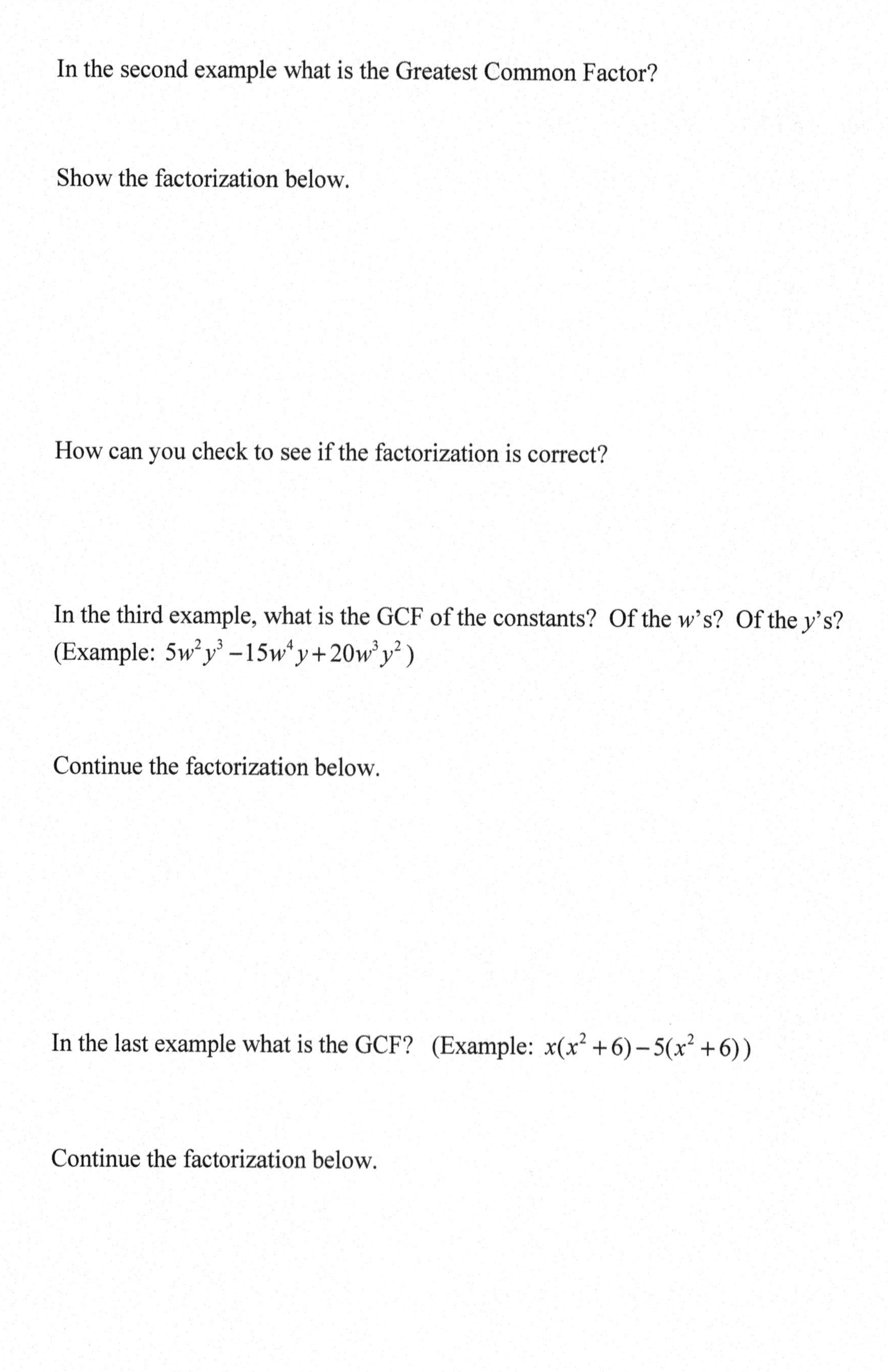

In the second example what is the Greatest Common Factor?

Show the factorization below.

How can you check to see if the factorization is correct?

In the third example, what is the GCF of the constants? Of the w's? Of the y's? (Example: $5w^2y^3 - 15w^4y + 20w^3y^2$)

Continue the factorization below.

In the last example what is the GCF? (Example: $x(x^2+6) - 5(x^2+6)$)

Continue the factorization below.

Work through Example 1 showing all work below.

Factor each polynomial by factoring out the GCF.

a. $5w^2y^3 - 15w^4y + 20w^3y^2$

b. $x(x^2 + 6) - 5(x^2 + 6)$

Section R.6 Objective 2: Factoring by Grouping.

Watch the video with Objective 2 and answer the questions below.

When is factoring by grouping a good thing to try?

In the first 4-term example, what are the two groups?

How is the first group factored? Show the factorization below.

How is the second group factored? Show the factorization below.

What is the common term of the two groups?

Write the complete factored form below.

In the next example, show the factorization step by step below.

Why was $-y$ the GCF of the second group?

How can you check your work?

Work the problems in Example 2 showing all steps below.

Factor each polynomial by grouping.

a. $6x^2 - 2x + 9x - 3$

b. $2x^2 + 6xw - xy - 3wy$

Section R.6 Objective 3: Factoring Trinomials with a Leading Coefficient Equal to One

Watch the video with Objective 3 and answer the following questions.

Use FOIL to multiply the two sets of two binomials below:

$(x+3)(x+5)$ $(x+3)(x-5)$

Factoring a Trinomial of the form x^2+bx+c,

1. If $c<0$, then try factors of the form ______________________.
2. If $c>0$, then look at the value of b.
    - If $b>0$, then try factors of the form ______________________.
    - If $b<0$, then try factors of the form ______________________.

In the first example in the video, we want to factor the trinomial $x^2-2x-24$.

What are the signs of the constants of the two binomial factors? Why?

What do you know about the product of the numbers in the two boxes?

What do you know about the sum of the numbers in the two boxes?

Write the factored form below.

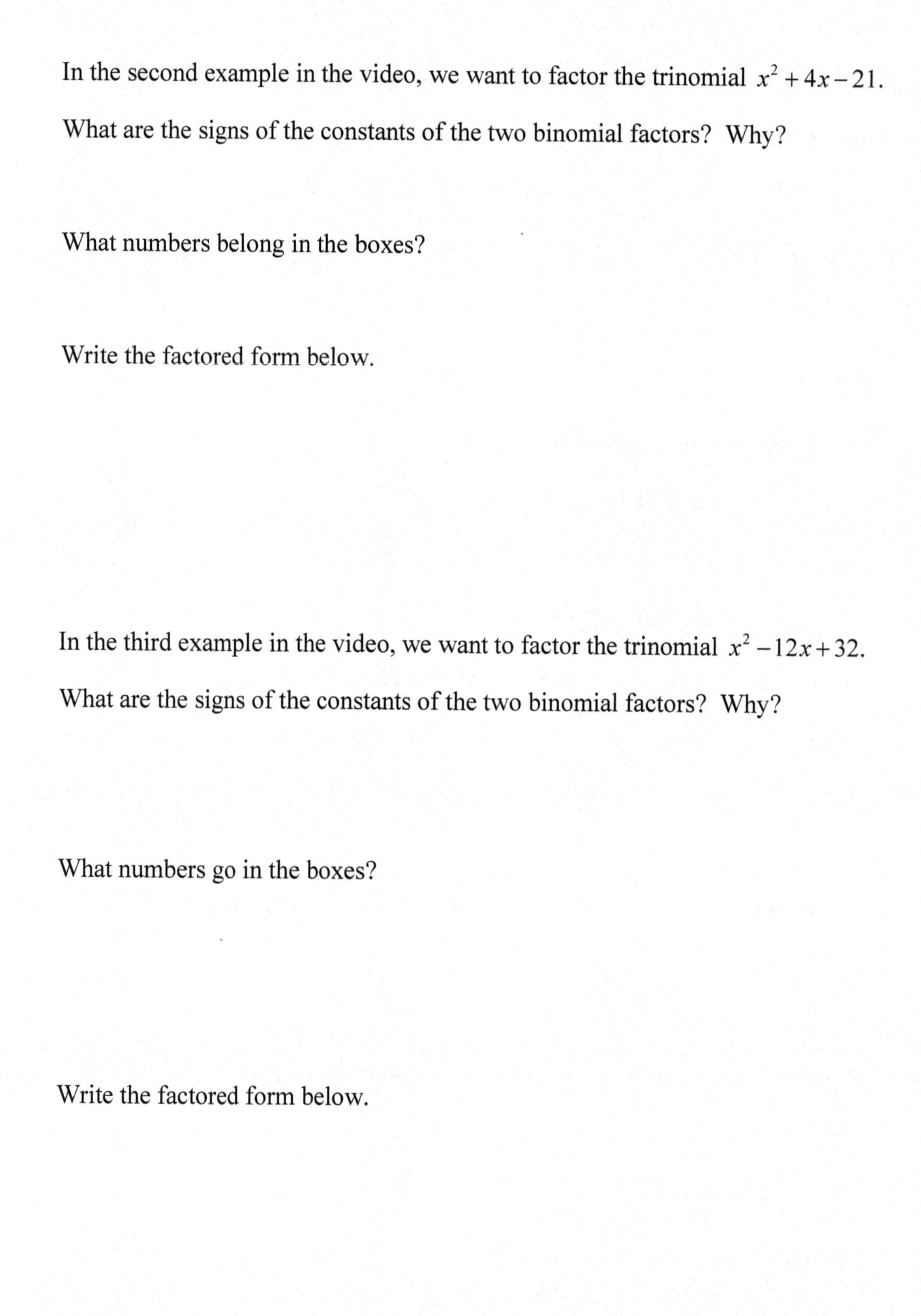

In the second example in the video, we want to factor the trinomial $x^2 + 4x - 21$.

What are the signs of the constants of the two binomial factors? Why?

What numbers belong in the boxes?

Write the factored form below.

In the third example in the video, we want to factor the trinomial $x^2 - 12x + 32$.

What are the signs of the constants of the two binomial factors? Why?

What numbers go in the boxes?

Write the factored form below.

Write the three steps for **Factoring Trinomials of the Form $x^2 + bx + c$.** (See page R.6-7.)

Step 1.

Step 2.

Step 3.

Use the three-step process seen above as you work through Example 3 showing all work below. Watch the video to check your work.

Factor each trinomial.

a. $x^2 - 2x - 24$

b. $x^2 - 2x - 6$

c. $x^2 - 12x + 32$

Section R.6 Objective 4: Factoring Trinomials with a Leading Coefficient Not Equal to One

Work through the animation seen toward the bottom of page R.6-10 and answer the following questions.

Write down the four steps for factoring trinomials of the form $ax^2 + bx + c,\ a \neq 1$ using the **Trial and Error Strategy** as seen in the animation.

Step 1:

Step 2:

Step 3:

Step 4:

In the animation, Click "CONTINUE" after the four steps are displayed.

Use the four steps above to factor the $6x^2 + 29x + 35$ trinomial.

Step 1: List all pairs of factors of 6.

Step 2: List all pairs of factors of 35.

Step 3: In the animation, click on the different combinations of the factors of 6 and the factors of 35 until you find the correct pair of binomial factors and complete the correct factorization:

$$6x^2 + 29x + 35 = (\_\_\_\_ + \_\_\_\_)(\_\_\_\_ + \_\_\_\_)$$

Step 4: Multiply the factors above to check your answer.

There is a method for factoring trinomials that does **not** involve the trial and error method. This method is known as the "*ac* method." Write down the five steps for this method as seen on page R.6-11.

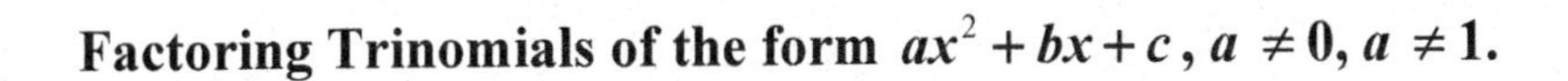

Factoring Trinomials of the form $ax^2 + bx + c$, $a \neq 0$, $a \neq 1$.

Step 1.

Step 2.

Step 3.

Step 4.

Step 5.

Work through Example 4 showing all steps below.

a. Factor the trinomial $4x^2 + 17x + 15$. Watch the animation to check your work.

b. Factor the trinomial $12x^2 - 25x + 12$.

Section R.6 Objective 5: Factoring Using Special Factoring Formulas

What is the **Difference of Two Squares**?

What does the tip say about the sum of two squares?

What are the 2 **perfect square trinomials**?

What are the formulas for a **sum or difference of 2 cubes**?

Work through the problems with Example 5. Watch the video to see if you got the correct solution.

Completely factor each expression.

a. $2x^3 + 5x^2 - 12x$

b. $5x^4 - 45x^2$

c. $8z^3x - 27y + 8z^3y - 27x$

Section R.7 Guided Notebook

Section R.7 Rational Expressions

- ☐ Work through Objective 1
- ☐ Work through Objective 2
- ☐ Work through Objective 3
- ☐ Work through Objective 4
- ☐ Work through Objective 5

Section R.7 Rational Expressions

Section R.7 Objective 1: Simplifying Rational Expressions

What is a **rational number**?

Fill in the blank: A rational expression is the quotient of two ________________ expressions.

Write the formal definition of a **Rational Expression**. Include 2 examples.

Work through Example 1 showing all work below. Watch the video to see a detailed explanation.

Simplify each rational expression.

a. $\dfrac{x^2+x-12}{x^2+9x+20}$

b. $\dfrac{x^3+1}{x+1}$

c. $\dfrac{x^2-x-2}{2x-x^2}$

Section R.7 Objective 2: Multiplying and Dividing Rational Expressions

How do you find the product of two rational expressions?

Work through Example 2 showing all steps below.

Multiply $\dfrac{2x^2+3x-2}{3x^2-2x-1}\cdot\dfrac{3x^2+4x+1}{2x^2+x-1}$

How do you find the quotient of two rational expressions?

Work through Example 3 and show work below.

Divide and simplify. $\dfrac{x^3-8}{2x^2-x-6} \div \dfrac{x^2+2x+4}{6x^2+11x+3}$

Work Example 4 and show all work below. Watch the video to see the complete solution and explanation.

Perform the indicated operations and simplify.

$$\frac{x^2+x-6}{x^2+x-42}\cdot\frac{x^2+12x+35}{x^2-x-2}\div\frac{x+7}{x^2+8x+7}$$

Section R.7 Objective 3: Adding and Subtracting Rational Expressions.

To add or subtract rational numbers they must have ______________________________.

Work through Example 5 showing all steps below.

Perform the indicated operations and simplify.

a. $\dfrac{3}{x+1}-\dfrac{2-x}{x+1}$

b. $\dfrac{3}{x^2+2x}+\dfrac{x-2}{x^2-x}$

Work through Example 6 answering the questions below. Then check your solutions by watching the video.

Answer the following questions for part a. $\frac{3}{x-y}-\frac{x+5y}{x^2-y^2}$

a. What is done first to find the LCD?

b. What form of 1 is used to modify the first fraction?

c. Continue simplifying the expression, showing all steps below.

Answer the following questions for part b. $\dfrac{x+4}{3x^2+20x+25}+\dfrac{x}{3x^2+16x+5}$

a. Show the steps to factor the first and second denominator.

b. What is the LCD?

c. What form of 1 is used to modify each fraction?

d. Show the remaining steps to simplify the expression.

Section R.7 Objective 4: Simplifying Complex Rational Expressions

What is a **complex rational expression**? Give 2 examples.

What are the 3 steps of **Method I for Simplifying Complex Rational Expressions**?

Work through Example 7 showing all steps below.

Simplify using Method I: $\dfrac{4-\dfrac{5}{x-1}}{\dfrac{6}{x-1}-7}$

What are the 3 steps of **Method II for Simplifying Complex Rational Expressions?**

Work through Example 8 and show all work below.

Simplify using Method II: $\dfrac{4-\dfrac{5}{x-1}}{\dfrac{6}{x-1}-7}$

Work Example 9 showing all steps below. Watch the video to see if you got the correct solution.

Simplify the complex rational expression using Method I or Method II.

$$\frac{-\frac{1}{x}-\frac{3}{x+4}}{\frac{2}{x^2+4x}+\frac{2}{x}}$$

Section R.7 Objective 5: Applications of Rational Expressions

Work through Example 10 showing all steps below. Watch the video to check your solution.

The area of a trapezoid can be represented by the expression $\frac{1}{2}(a+b)h$, where a and b represent the lengths of the parallel bases of the trapezoid and h represents the height. Suppose that $a=\frac{1}{x}$ inches, $b=\frac{1}{x+1}$ inches, and $h=x^2-x$ inches.

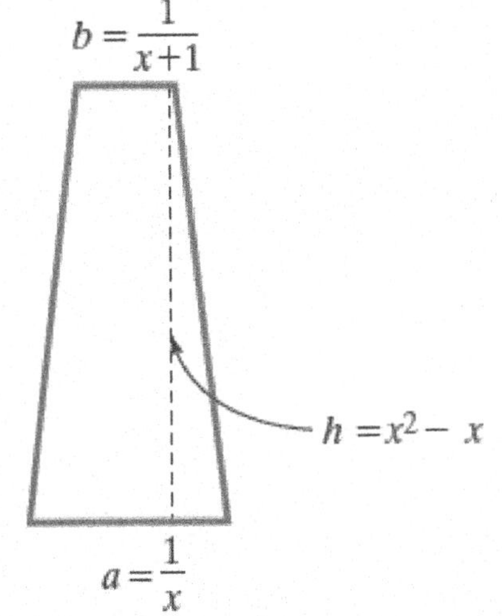

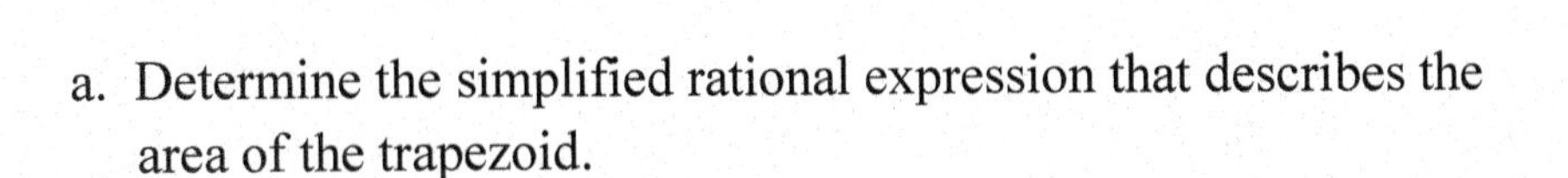

a. Determine the simplified rational expression that describes the area of the trapezoid.

b. What is area of the trapezoid if $x=2$ inches?

Section 1.1 Guided Notebook

Section 1.1 Linear Equations

- ▯ Work through Section 1.1 TTK #1
- ▯ Work through Section 1.1 TTK #2
- ▯ Work through Section 1.1 TTK #3
- ▯ Work through Objective 1
- ▯ Work through Objective 2
- ▯ Work through Objective 3
- ▯ Work through Objective 4
- ▯ Work through Objective 5

Section 1.1 Linear Equations

Section 1.1 Objective 1 Recognizing Linear Equations

What is the definition of an **algebraic expression**?

What is the definition of a **linear equation in one variable**?

In the Interactive Video following the definition of a linear equation in one variable, which equation is not linear? Explain why it is not linear.

Section 1.1 Objective 2 Solving Linear Equations with Integer Coefficients

What does the term **integer coefficient** mean?

Work through Example 1 and take notes here.
Solve $5(x-6)-2x=3-(x+1)$.

Work through Example 2 and take notes here. Watch the video to check your solution.
Solve $6-4(x+4)=8x-2(3x+5)$.

Section 1.1 Objective 3 Solving Linear Equations Involving Fractions

What is the definition of a **least common denominator (LCD)**?

What is the first thing to do when solving linear equations involving fractions?

Work through the video that accompanies Example 3 and write your notes here:

Solve $\frac{1}{3}(1-x)-\frac{x+1}{2}=-2$

Section 1.1 Objective 4 Solving Linear Equations Involving Decimals

When encountering a linear equation involving decimals, how do you eliminate the decimals?

Work through the video that accompanies Example 4 and write your notes here:
Solve $0.1(y-2)+0.03(y-4)=0.02(10)$.

Section 1.1 Objective 5 Recognizing Rational Equations

What is a **rational number**?

What is a **rational expression**?

Write down the definition of a **rational equation** and write down at least one example of a rational equation.

Work through Example 5. Determine which of the following equations are rational equations. (Watch the video to check to see if you are correct.)

a. $\frac{2-x}{x+5}+3=\frac{4}{x+2}$ b. $x^2-2x-24=\frac{1}{2}$ c. $\frac{12}{x^2+x-2}-\frac{x+3}{x-1}=\frac{1-x}{x+2}$

Section 1.1 Objective 6 Solving Rational Equations that Lead to Linear Equations

Fill in the blanks below:

The process of solving a rational equation is very similar to the process of solving linear equations containing fractions. That is, we first determine the

____

and then we ______________________ both sides of the equation by the _______________.

We have to be extra cautious when solving rational equations because we have to be aware of

________________________________.

What is a **restricted value?**

Work through Example 6 and take notes here: $\frac{2-x}{x+2}+3=\frac{4}{x+2}$

What is the definition of an **extraneous solution**?

Explain why there was an extraneous solution to Example 6.

Write down the five steps for **Solving Rational Equations.**

Step 1

Step 2

Step 3

Step 4

Step 5

Work through Example 7 by following the five steps above and take notes here. Watch the video to check your solution.

Solve $\frac{2}{x+4}+\frac{1}{x-5}=\frac{5}{x^2-x-20}$.

Work through Example 8 by following the five steps for solving rational equations and take notes here. Watch the video to check your solution.

Solve $\dfrac{12}{x^2+x-2}-\dfrac{x+3}{x-1}=\dfrac{1-x}{x+2}$.

Section 1.2 Guided Notebook

Section 1.2 Applications of Linear and Rational Equations

- ☐ Work through TTK #4
- ☐ Work through Objective 1
- ☐ Work through Objective 2
- ☐ Work through Objective 3
- ☐ Work through Objective 4
- ☐ Work through Objective 5

Section 1.2 Applications of Linear and Rational Equations

Section 1.2 Objective 1: Converting Verbal Statements into Mathematical Statements

Work through Example 1 and show all work below.
Watch the video to check your solutions.

Rewrite each statement as an algebraic expression or equation.

a. 7 more than three times a number.

b. 5 less than twice a number

c. Three times the quotient of a number and 11

d. The sum of a number and 9 is 1 less than half of the number.

e. The product of a number and 4 is 1 more than 8 times the difference of 10 and the number.

What is the caution statement with Example 1 part b about?

Who is George Polya and what is he known for?

Write down Polya's Guidelines for Problem Solving.

What is the Four-Step Strategy for Problem Solving?

Step 1

Step 2

Step 3

Step 4

Section 1.2 Objective 2: Solving Applications Involving Unknown Numeric Quantities

Work through Example 2 and show all work below.

Roger Staubach and Terry Bradshaw were both quarterbacks in the National Football League. In 1973, Staubach threw three touchdown passes more than twice the number of touchdown passes thrown by Bradshaw. If the total number of touchdown passes between Staubach and Bradshaw was 33, how many touchdown passes did each player throw?

Work through Example 3 and show your work below. Watch the video to check your solution.

One number is three times another number. Determine the two numbers if the sum of their reciprocals is 4.

Section 1.2 Objective 3: Solving Applications Involving Geometric Formulas

Work through Example 4 and show your work below. Watch the video to check your solution.

The length of a college basketball court (rectangle) is 6 feet less than twice its width. If the perimeter is 288 feet, what are the dimensions of the court?

Work through Example 5 and show your work below. Watch the video to check your solution.

Suppose that the area of a trapezoid can be represented by the expression $\frac{6}{x}$ square feet and suppose that the lengths of the parallel bases of the trapezoid can be represented by the expressions $\frac{2}{x}$ feet and $\frac{3}{x+1}$ feet respectively. If the height is 4 feet, determine the value of x. Then determine the area and values of the lengths of the parallel bases.

Section 1.2 Objective 4: Solving Applications Involving Decimal Equations (Money, Mixture, Interest)

Work through the video with Example 6 and take notes below.

Billy has $16.50 in his piggy bank, consisting of nickels, dimes, and quarters. Billy notices that he has 20 fewer quarters than dimes. If the number of nickels is equal to the number of quarters and dimes combined, how many of each coin does Billy have?

Work through the video with Example 7 and show all work below.

How many milliliters of a 70% acid solution must be mixed with 30 mL of a 40% acid solution to obtain a mixture that is 50% acid?

Work through the video with Example 8 and show all work below.

Kristen inherited $20,000 from her aunt Dawn Ann, with the stipulation that she invest part of the money in an account paying 4.5% simple interest and the rest in an account paying 6% simple interest locked in for 3 years. If at the end of 1 year, the total interest earned was $982.50, how much was invested at each rate?

Section 1.3 Guided Notebook

Section 1.3 Complex Numbers

- ☐ Work through Section 1.3 TTK
- ☐ Work through Objective 1
- ☐ Work through Objective 2
- ☐ Work through Objective 3
- ☐ Work through Objective 4
- ☐ Work through Objective 5

Section 1.3 Complex Numbers

THE IMAGINARY UNIT

Take notes on the video that explains the imaginary unit here:

What is the definition of the **imaginary unit**?

Section 1.3 Objective 1 Simplifying Powers of i

Explain the cyclic nature of powers of i:

Work through Example 1 and take notes here:

Simplify each of the following:

a. i^{43} b. i^{100} c. i^{-21}

COMPLEX NUMBERS

What is a complex number?

Give several examples of complex numbers.

Is every real number considered a complex number? Why or why not?

Section 1.3 Objective 2 Adding and Subtracting Complex Numbers

Watch the video, work through Example 2 and explain how to add/subtract complex numbers.

Perform the indicated operations:

a. $(7-5i)+(-2-i)$

b. $(7-5i)-(-2-i)$

Section 1.3 Objective 3 Multiplying Complex Numbers

Fill in the blanks:

When multiplying complex numbers, treat the problem as if it were the multiplication of two ______________________. Just remember that _____ = ______ .

Work through Example 3 and show your work here. Watch the video to check your solution.

Multiply $(4-3i)(7+5i)$.

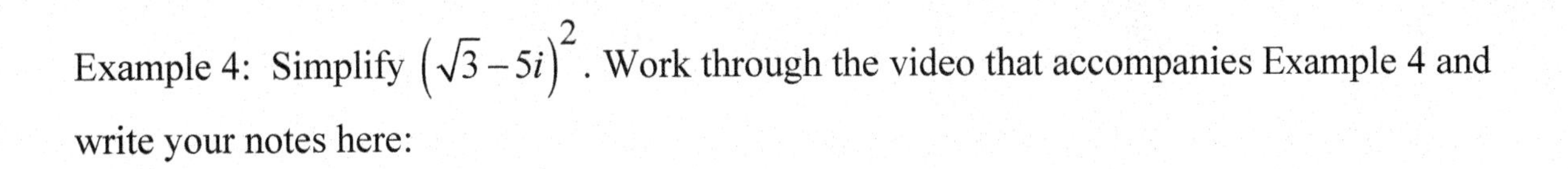

Example 4: Simplify $\left(\sqrt{3}-5i\right)^2$. Work through the video that accompanies Example 4 and write your notes here:

What is the definition of a **complex conjugate**?

Work through Example 5 and show your work:
Multiply the complex number $z=-2-7i$ by its complex conjugate $\overline{z}=-2+7i$.

What will **always** happen when you multiply a complex number by its complex conjugate?

Section 1.3 Objective 4 Finding the Quotient of Complex Numbers

What is the goal when dividing two complex numbers?

Work through Example 6 and take notes here. Watch the video at the top of page 1.3-11 to check your solution.

Write the quotient in the form $a+bi$: $\dfrac{1-3i}{5-2i}$

Section 1.3 Objective 5 Simplifying Radicals with Negative Radicands

Write down the property seen on page 1.3-12.

Work through Example 7 and write your notes here:

Simplify: $\sqrt{-108}$

True or False: $\sqrt{a}\sqrt{b} = \sqrt{ab}$ for all real numbers a and b. Explain.

Work through Example 8 and write your notes here: Simplify the following expressions:

a) $\sqrt{-8} + \sqrt{-18}$

b) $\sqrt{-8} \cdot \sqrt{-18}$

c) $\dfrac{-6 + \sqrt{(-6)^2 - 4(2)(5)}}{2}$

d) $\dfrac{4 \pm \sqrt{-12}}{4}$

Section 1.4 Guided Notebook

Section 1.4 Quadratic Equations

- ☐ Work through Section 1.4 TTK
- ☐ Work through Objective 1
- ☐ Work through Objective 2
- ☐ Work through Objective 3
- ☐ Work through Objective 4
- ☐ Work through Objective 5

Section 1.4 Quadratic Equations

What is the definition of a **quadratic equation in one variable**?

Section 1.4 Objective 1 Solving Quadratic Equations by Factoring and the Zero Product Property

Watch the video located under Objective 1 and take notes here:
(Be sure that you write down and understand the **zero product property.**)

Work through Example 1. Watch the video to check your solution.
Solve $6x^2 - 17x = -12$.

Section 1.4 Objective 2 Solving Quadratic Equations Using the Square Root Property

Watch the video located just under Objective 2 and take notes on this page:

What is the square root property and when can we use it when solving quadratic equations?

Work through Example 2 in your eText (as seen in the video) and take notes here:

a) $x^2 - 16 = 0$

b) $2x^2 + 72 = 0$

c) $(x-1)^2 = 7$

Section 1.4 Objective 3 Solving Quadratic Equations by Completing the Square

Write down three perfect square trinomials and factor each as a binomial squared.

What is the relationship between the linear term (x-term) and the constant term of every perfect square trinomial?

Work through Example 3 and take notes here.
What number should be added to each binomial to make it a perfect square trinomial?

a) $x^2 - 12x$

b) $x^2 + 5x$

c) $x^2 - \frac{3}{2}x$

Write down the **5 Steps for Solving $ax^2 + bx + c = 0$, $a \neq 0$ by Completing the Square.**

1.

2.

3.

4.

5.

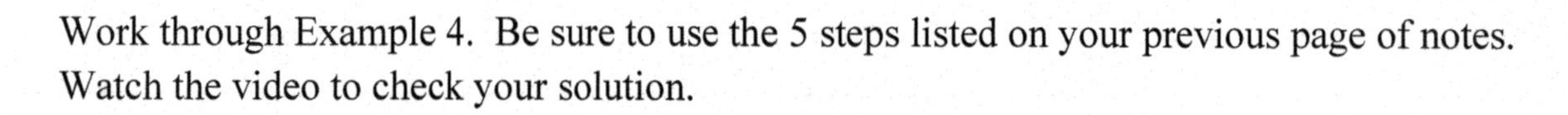

Work through Example 4. Be sure to use the 5 steps listed on your previous page of notes. Watch the video to check your solution.

Solve $3x^2 - 18x + 19 = 0$ by completing the square.

Work through Example 5. Be sure to use the 5 steps listed on your previous page of notes:

Solve $2x^2 - 10x - 6 = 0$ by completing the square.

Section 1.4 Objective 4 Solving Quadratic Equations Using the Quadratic Formula

You have all probably seen the quadratic formula, but where does it come from? Work through the animation that shows how to derive the quadratic formula and take notes here:

Deriving the Quadratic Formula

Start with the equation $ax^2 + bx + c = 0,\ a \neq 0$ and solve for x.

Step 1.

Step 2.

Step 3.

Step 4.

Step 5.

Write down the **Quadratic Formula:**

Work through the Example 6 video and write your notes here.
Solve $3x^2 + 2x - 2 = 0$ using the quadratic formula.

Work through the Example 7 video and write your notes here.
Solve $4x^2 - x + 6 = 0$ using the quadratic formula.

Section 1.4 Objective 5 Using the Discriminant to Determine the Type of Solutions of a Quadratic Equation

Watch the video located under Objective 5 and take notes here:

Work through Example 8 and take notes here. Watch the video to check your solutions.

Use the discriminant to determine the number and nature of the solutions to each of the following quadratic equations:

a) $3x^2 + 2x + 2 = 0$

b) $4x^2 + 1 = 4x$

Section 1.5 Guided Notebook

Section 1.5 Applications of Quadratic Equations

- [ ] Work through Objective 1
- [ ] Work through Objective 2
- [ ] Work through Objective 3
- [ ] Work through Objective 4
- [ ] Work through Objective 5

Section 1.5 Applications of Quadratic Equations

Write down the **Four-Step Strategy for Problem Solving.**

Section 1.5 Objective 1 Solving Applications Involving Unknown Numeric Quantities

Answer the following question for Example 1.

The product of a number and 1 more than twice the number is 36. Find the two numbers.

Explain why the other number is represented by $2x+1$

Explain why the equation is $x(2x+1)=36$.

Show all steps to solve the equation below.

Show the steps to substitute $-\frac{9}{2}$ for x in $2x+1$.

What is the answer to the question in Example 1?

Section 1.5 Objective 2 Using the Projectile Motion Model

What is the projectile motion model seen in this objective?

Work through the video that accompanies Example 2 and take notes here:

A toy rocket is launched at an initial velocity of 14.7 m/s from a 49-m tall platform. The height h of the object at any time t seconds after launch is given by the equation $h = -4.9t^2 + 14.7t + 49$. When will the rocket hit the ground?

Another model used to describe projectile motion (where the height is in feet and time is in seconds) is given by $h = -16t^2 + v_0 t + h_0$ (where v_0 is the initial velocity and h_0 is initial height above the ground).

Section 1.5 Objective 3 Solving Geometric Applications

Work through the interactive video that accompanies Example 3 and write your notes here: The length of a rectangle is 6 in. less than four times the width. Find the dimensions of the rectangle if the area of the rectangle is 54 in^2.

Work through the video that accompanies Example 4 and take notes here:
Jimmy bought a new 40 in. high-definition television. If the length of Jimmy's television is 8 in. longer than the width, find the width of the television. (Remember the Pythagorean Theorem: $a^2 + b^2 = c^2$)

Section 1.6 Guided Notebook

Section 1.6 Other Types of Equations

- ▯ Work through Section 1.6 TTK #1
- ▯ Work through Section 1.6 TTK #2
- ▯ Work through Objective 1
- ▯ Work through Objective 2
- ▯ Work through Objective 3

Section 1.6 Other Types of Equations

Section 1.6 Objective 1 Solving Higher-Order Polynomial Equations

Work through Example 1 and take notes here. Watch the video to check your solutions.

Find all solutions of the equation $3x^3 - 2x = -5x^2$.

Work through Example 2 and take notes here. Watch the video to check your solutions.
Find all solutions of the equation $2x^3 - x^2 + 8x - 4 = 0$.

Section 1.6 Objective 2 Solving Equations That are Quadratic In Form (Disguised Quadratics)

What does it mean for an equation to be "quadratic in form?"

Work through the interactive video that accompanies Example 3 and solve each equation:

Example 3a: $2x^4 - 11x^2 + 12 = 0$

Example 3b: $\left(\frac{1}{x-2}\right)^2 + \frac{2}{x-2} - 15 = 0$

Example 3c: $x^{2/3} - 9x^{1/3} + 8 = 0$ (Hint: $\left(x^a\right)^b = x^{ab}$)

Example 3d: $3x^{-2} - 5x^{-1} - 2 = 0$

Section 1.6 Objective 3 Solving Equations Involving Radicals

Work through Example 4 taking notes here. Watch the video to check your work.

Solve $\sqrt{x-1} - 2 = x - 9$.

As indicated in the video, make sure that you *always* isolate the radical prior to squaring both sides of an equation that involves a square root.

What is an **extraneous solution**?

Why is it important to check your solutions when solving equations involving radicals?

Work through the video that accompanies Example 5 taking notes here:
Solve $\sqrt{2x+3}+\sqrt{x-2}=4$.

Work through the video that accompanies Example 6 taking notes here:
Solve $\sqrt[3]{1-4x}+3=0$.

Section 1.7 Guided Notebook

Section 1.7 Linear Inequalities

- [ ] Work through Section 1.7 TTK #1
- [ ] Work through Section 1.7 TTK #2
- [ ] Work through Objective 1
- [ ] Work through Objective 2
- [ ] Work through Objective 3
- [ ] Work through Objective 4

Section 1.7 Linear Inequalities in One Variable

Read through the Introduction to Section 1.7 and answer the questions below:

Consider the inequality $2x-3\leq 5$. Explain why the numbers $-1, \frac{7}{2}$ and 4 are all solutions to this inequality.

There are acutely infinitely many solutions to the inequality $2x-3\leq 5$. List three different ways that are typically used to describe the solution to an inequality.

1.

2.

3.

Section 1.7 Objective 1 Solving Linear Inequalities

What is the definition of a **linear inequality in one variable**?

Be sure that you are familiar with the properties of linear inequalities that are discussed in the following table.

Properties of Inequalities

Let *a, b,* and *c* be real numbers:

	Property	In Words	Example
1	If $a < b$, then $a + c < b + c$	The same number may be added to both sides of an inequality.	$-3 < 7$ $-3 + 4 < 7 + 4$ $1 < 11$
2	If $a < b$, then $a - c < b - c$	The same number may be subtracted from both sides of an inequality.	$9 \geq 2$ $9 - 6 \geq 2 - 6$ $3 \geq -4$
3	For $c > 0$, if $a < b$, then $ac < bc$	Multiplying both sides of an inequality by a *positive* number *does not* switch *the direction* of the inequality.	$3 > 2$ $(3)(5) > (2)(5)$ $15 > 10$
4	For $c < 0$, if $a < b$, then $ac > bc$	Multiplying both sides of an inequality by a *negative* number switches *the direction* of the inequality.	$3 > 2$ $(3)(-5) < (2)(-5)$ $-15 < -10$
5	For $c > 0$, if $a < b$, then $\frac{a}{c} < \frac{b}{c}$	Dividing both sides of an inequality by a *positive* number *does not* switch *the direction* of the inequality.	$6 > 4$ $\frac{6}{2} > \frac{4}{2}$ $3 > 2$
6	For $c < 0$, if $a < b$, then $\frac{a}{c} > \frac{b}{c}$	Dividing both sides of an inequality by a *negative* number switches *the direction* of the inequality.	$6 > 4$ $\frac{6}{-2} < \frac{4}{-2}$ $-3 < -2$

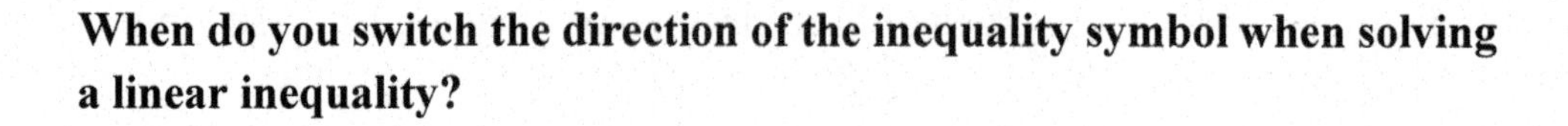

When do you switch the direction of the inequality symbol when solving a linear inequality?

Watch the **animation** seen on page 1.7-7 to see why it is important to be very careful when working with inequalities. Take your animation notes here.

Work through Example 1 and take notes here. Watch the video to check your solution.

Solve the inequality $2-5(x-2)<4(3-2x)+7$.
Express your answer in set-builder notation.

Work through Example 2 and take notes here. Watch the video to check your solution.

Solve the inequality $-9x-3\geq 7-4x$. Graph the solution set on a number line, and express the answer in interval notation.

Work through Example 3 and take notes here. Watch the interactive video to check your solutions.

Solve the linear inequalities.

a. $\frac{1-4w}{5}-\frac{w}{2}\leq -5$

b. $3.7y-6>6.1+3.45y$

Section 1.7 Objective 2 Solving Three-Part Inequalities in One Variable

Work through Example 4 taking notes here. Watch the video to check your solution.

Solve the inequality $-2 \le \frac{2-4x}{3} < 5$. Graph the solution set on a number line, and write the solution in set-builder notation.

Section 1.7 Objective 3 Solving Compound Inequalities in One Variable

Fill in the blanks:

A **compound inequality** consists of two inequalities that are joined together using the words ________ or _______ .

A number is a solution to a compound inequality involving the word "and" if that number is a solution to _________ inequalities.

A number is a solution to a compound inequality involving the word "or" if that number is a solution to ________________ inequality.

Write down the **Guidelines for Solving Compound Linear Inequalities.**

Step 1.

Step 2.

Step 3.

Work through Example 5 and take notes here. Watch the video to check your solution.

Solve $2x-7<-1$ *and* $3x+5\geq 3$. Write the solution in interval notation.

Work through Example 6 and take notes here. Watch the video to check your solution.

Solve $1-3x \geq 7$ *or* $3x+4>7$. Write the solution in interval notation.

Work through Example 7 and take notes here. Watch the video to check your solution.

Solve $3x-1<-7$ *and* $4x+1>9$.

Section 1.7 Objective 4 Solving Linear Inequality Word Problems

Write down the 6-step **Strategy for Solving Application Problems Involving Linear Inequalities**

Step 1.

Step 2.

Step 3.

Step 4.

Step 5.

Step 6.

Work through Example 8 and take notes here:

Suppose you rented a forklift to move a pallet with 70-lb blocks stacked on it. The forklift can carry a maximum of 2,535 lbs. If the pallet weighs 50 lbs. by itself with no blocks, how many blocks can be stacked on a pallet and lifted by the forklift?

Work through Example 9 and take notes here. Watch through the video to check your solution.

The perimeter of a rectangular fence is to be at least 80 feet and no more than 140 feet. If the width of the fence is 12 feet, what is the range of values for the length of the fence?

Work through Example 10 and take notes here. Watch through the video to check your solution.

Suppose that a wireless phone company offers a monthly plan for a smartphone that includes 4 GB of data for $110. Each additional GB of data (or fraction thereof) costs $15. If Antoine subscribes to this plan, how many GB of data can he use each month while keeping his total monthly cost to no more than $180 (before taxes)?

Work through Example 11 and take notes here. Watch through the video to check your solution.

An online retailer sells plush toys. She purchases the toys at the wholesale price of \$2.75 each and sells them online for \$7.75. If her fixed costs are \$900, how many plush toys must she sell in order to make a profit? Solve the inequality $R > C$ with R as her revenue and C as her cost.

Section 1.8 Guided Notebook

Section 1.8 Absolute Value Equations and Inequalities

- ☐ Work through Section 1.8 TTK
- ☐ Work through Section 1.8 TTK
- ☐ Work through Objective 1
- ☐ Work through Objective 2
- ☐ Work through Objective 3

Section 1.8 Absolute Value Equations and Inequalities

An Introduction to Absolute Value

Read the introduction to Section 1.8 and watch the video that accompanies it and take notes here:

Complete **Table 1** below:

Table 1 Absolute Value Equations and Inequality Properties

Let u be an algebraic expression and c be a real number such that $c > 0$, then

1. $|u| = c$ equivalent to ______________________________.
2. $|u| < c$ equivalent to ______________________________.
3. $|u| > c$ equivalent to ______________________________.

Section 1.8 Objective 1 Solving an Absolute Value Equation

Work through the video that accompanies Example 1 and take notes here:

Solve $|1-3x| = 4$.

Note that the absolute value equation in Example 1 is in *standard form.* Make sure that you *always* get your absolute value equations into *standard form* before solving the equation.

Work through the interactive video that accompanies Example 2 and take notes here:

Solve each absolute value equation.

a. $-2|1-3x|+5=-9$

b. $\left|\frac{5x-1}{x+3}\right|=11$

c. $|x^2-x-8|=4$

d. $|\sqrt{2x+9}-x|=3$

Section 1.8 Objective 2 Solving an Absolute Value Inequalities

Fill in the blanks.

When we encounter an inequality of the form $|u| < c,$ where c is a positive constant, property _____ states that the inequality is equivalent to the three-part inequality ______________________________.

Work through the video that accompanies Example 3 and take notes here:

Solve $|4x-3|+2 \leq 7$.

Fill in the blanks.

When we encounter an inequality of the form $|u| > c$, where c is a positive constant, property _____ states that the inequality is equivalent to the compound inequality ______________________________.

Work through the video that accompanies Example 4 and take notes here:

Solve $|5x+1| > 3$.

Work through the video that accompanies Example 5 and take notes here:

Solve $7-2|1-4x| \leq -9$.

Work through the interactive video that accompanies Example 6 and take notes here:
Solve each of the following.

a. $|3x-2|=0$

b. $|x+6|=-4$

c. $|7x+5|\leq 0$

d. $|3-4x|<-6$

e. $|8x-3|>0$

f. $|1-9x|\geq -5$

Section 1.9 Guided Notebook

Section 1.9 Polynomial and Rational Inequalities

- ☐ Work through Section 1.9 TTK #1
- ☐ Work through Section 1.9 TTK #2
- ☐ Work through Section 1.9 TTK #3
- ☐ Work through Section 1.9 TTK #4
- ☐ Work through Objective 1
- ☐ Work through Objective 2

Section 1.9 Polynomial and Rational Inequalities

Section 1.9 Objective 1 Solving Polynomial Inequalities

Watch the video that accompanies Example 1 and write your notes here:
Solve the polynomial inequality $x^3 - 3x^2 + 2x \geq 0$. (Note that there is room on the next page to write down the 7 steps for solving polynomial inequalities.)

What is a **boundary point**?

What is a **test value**?

Write down the 7 Steps for Solving Polynomial Inequalities

1.

2.

3.

4.

5.

6.

7.

Follow these 7 steps to solve the polynomial inequality in Example 2:
Solve $x^2 + 5x < 3 - x^2$.

Section 1.9 Objective 2 Solving Rational Inequalities

What is the definition of a **rational inequality**?

Write down the 7 Steps for **Solving Rational Inequalities**.

1.

2.

3.

4.

5.

6.

7.

Watch the video that accompanies Example 3 and write your notes here:

Solve $\frac{x-4}{x+1} \geq 0$.

(Use the 7-step method for solving rational inequalities that are on your previous page of notes. Pay close attention to the boundary point obtained from the denominator. How will you plot this boundary point on a number line?)

Watch the video that accompanies Example 4 and write your notes here:

Solve $x > \dfrac{3}{x-2}$.

Section 2.1 Guided Notebook

Section 2.1 The Rectangular Coordinate System

- ☐ Work through Section 2.1 TTK #1
- ☐ Work through Section 2.1 TTK #2
- ☐ Work through Section 2.1 TTK #3
- ☐ Work through Section 2.1 TTK #4
- ☐ Work through Objective 1
- ☐ Work through Objective 2
- ☐ Work through Objective 3
- ☐ Work through Objective 4

Section 2.1 The Rectangular Coordinate System

Read the introduction to Section 2.1 and write notes here:

Watch the animation seen on page 2.1-6. Draw a rectangular coordinate system and label the four quadrants. Then, plot the two ordered pairs seen in the animation.

Section 2.1 Objective 1 Plotting Ordered Pairs

Work through the video that accompanies Example 1 and write your notes here:

Plot the ordered pairs $(-2,3),(0,4),(2,5),$ and $(4,6)$ and state in which quadrant or on which axis each pair lies.

What is the equation of the graph of the straight line that passes through the 4 ordered pairs from Example 1?

Section 2.1 Objective 2 Graphing Equations by Plotting Points

Work through the video that accompanies Example 2 and write your notes here:

Sketch the graph of $y = x^2 - 4x + 4$.

Work through the video that accompanies Example 3 and write your notes here:
Determine whether the following ordered pairs lie on the graph of the equation $x^2 + y^2 = 1$.

a. $(0, -1)$

b. $(1, 0)$

c. $\left(\frac{1}{3}, \frac{2}{3}\right)$

d. $\left(-\frac{\sqrt{2}}{2}, \frac{\sqrt{2}}{2}\right)$

Section 2.1 Objective 3 Finding Intercepts of a Graph Given an Equation

What is the definition of the **intercepts of a graph**?

Fill in the blanks:
A y-intercept is the ____________________ of a point where a graph touches or crosses the ______________.

An x-intercept is the ____________________ of a point where a graph touches or crosses the ______________.

Complete the sentences below that describe how to algebraically find x- and y-intercepts.

Algebraically Finding x- and y-Intercepts Given an Equation in Two Variables

Finding x-intercepts: Set all values of the variable ________________________________.

Finding y-intercepts: Set all values of the variable ________________________________.

Work through the video that accompanies Example 4 and write your notes here:
Find the x- and y-intercepts of the graphs of the given equations.

a. $y=\dfrac{2x-1}{x+3}$

b. $\sqrt{x+2}+y=3$

c. $(x-1)^2+(y-3)^2=5$

Section 2.1 Objective 4 Finding the Midpoint of a Line Segment Using the Midpoint Formula

Write down the **midpoint formula** here:

Work through the video that accompanies Example 5 and write your notes here:
Find the midpoint of the line segment whose endpoints are $(-3,2)$ and $(4,6)$.

Section 2.1 Objective 5 Finding the Distance Between Two Points Using the Distance Formula

Watch the video that accompanies Objective 5. Take notes below.

Write the **distance formula** here:

Work through the video that accompanies Example 7 and write your notes here:

Find the distance between the points $A(-1,5)$ and $B(4,-5)$.

Section 2.2 Guided Notebook

Section 2.2 Circles

- [ ] Work through Section 2.2 TTK #1
- [ ] Work through Section 2.2 TTK #3
- [ ] Work through Section 2.2 TTK #4
- [ ] Work through Objective 1
- [ ] Work through Objective 2
- [ ] Work through Objective 3

Section 2.2 Circles

An Introduction to Circles
Read the introduction to Section 2.2.

What is the definition of a circle?

Show how to derive the equation of a circle by watching the **animation** found in the introduction and take notes here:

Write the **standard form of an equation of a circle** here:

Section 2.2 Objective 1 Writing the Standard Form of an Equation of a Circle

Work through the video that accompanies Example 1 and take notes here:

Find the standard form of the equation of the circle whose center is $(-2,3)$ and with radius 6.

Work through Example 2 and take notes here:

Find the standard form of a circle whose center is $(0,6)$ and that passes through the point $(4,2)$.

Work through the video that accompanies Example 3 and take notes here:
Find the standard form of the equation of the circle that contains endpoints of a diameter at $(-4,-3)$ and $(2,-1)$.

Section 2.2 Objective 2 Sketching the Graph of a Circle

Work through the **Guided Visualization** titled "Sketching the Graph of a Circle" seen on page 2.2-7. Experiment by changing the values of the coordinates of the center and the radius. Then sketch two circles below. One circle should have a center located in Quadrant I and the other circle should have a center located in Quadrant IV.

Work through the video that accompanies Example 4 and take notes here:
Find the center and radius, and sketch the graph of the circle $(x-1)^2+(y+2)^2=9$.
Also find any intercepts. **(Be sure to pay close attention to how we find intercepts.)**

Note: When determining intercepts algebraically, what does it mean if you get an imaginary solution?

Section 2.2 Objective 3 Converting the General Form of a Circle into Standard Form

What is the difference between standard form and general form?

Work through the video that accompanies Example 5 and take notes here:
Write the equation $x^2+y^2-8x+6y+16=0$ in standard form; find the center, radius, and intercepts, and sketch the graph.

Carefully work through the animation that accompanies Example 6 and take notes here:
Write the equation $4x^2 + 4y^2 + 4x - 8y + 1 = 0$ in standard form; find the center, radius, and intercepts, and sketch the graph.

Section 2.3 Guided Notebook

Section 2.3 Lines

- ▯ Work through Objective 1
- ▯ Work through Objective 2
- ▯ Work through Objective 3
- ▯ Work through Objective 4
- ▯ Work through Objective 5
- ▯ Work through Objective 6
- ▯ Work through Objective 7
- ▯ Work through Objective 8

Section 2.3 Lines

<u>Section 2.3 Objective 1 Determining the Slope of a Line</u>

Watch the video that accompanies Objective 1.

Write down the **definition of slope** here:

Work through the **Guided Visualization** titled "Determining the Slope of a Line" found on the bottom of page 2.3-4. Then draw 4 lines below. One line should have positive slope, one line should have negative slope, one line should have zero slope, and one line should have undefined slope.

Work through Example 1: Find the slope of the line that passes through the indicated ordered pairs.

a. $(6,-4)$ and $(-5,1)$

b. $(3,-1)$ and $(3,6)$

c. $(-5,4)$ and $(-2,4)$

Section 2.3 Objective 2 Sketching a Line Given a Point and the Slope

Work through the video that accompanies Example 2 and take notes here:

Sketch the line with slope $m=\frac{2}{3}$ that passes through the point $(-1,-4)$.

Also, find three more points located on the line.

Section 2.3 Objective 3 Finding the Equation of a Line Using the Point-Slope Form

Watch the video that accompanies Objective 3 to see how to derive the Point-Slope Form of the equation of a line and take notes here:

Write down the **Point-Slope Form** of a line here:

Work through Example 3 and take notes here:

Find an equation in point-slope form of the line with slope $m=\frac{2}{3}$ that passes through the point $(-1,-4)$.

Work through the **Guided Visualization** titled "Equations of Lines: Point-Slope" seen on the bottom of page 2.3-9. Draw a line below that passes through the point $(-1,-4)$ with slope $m=-2$. According to the **Guided Visualization,** what is the equation of that line?

Section 2.3 Objective 4 Finding the Equation of a Line Using the Slope-Intercept Form

In Example 3 from the previous page, you should have found the equation of the line to be $y+4=\frac{2}{3}(x+1)$. Try solving this equation for y. What do you get?

Now, write down the **Slope-Intercept Form** of the equation of a line here:

Work through the video that accompanies Example 4 and take notes here:

Find the equation of the line with slope $\frac{1}{4}$ and y-intercept 3, and write your answer in slope-intercept form.

Work through the **Guided Visualization** titled "Equations of Lines: Slope-Intercept" seen on the bottom of page 2.3-14. Draw a line below that passes through the point $(0,-4)$ with slope $m=-3$. According the **Guided Visualization,** what is the equation of that line?

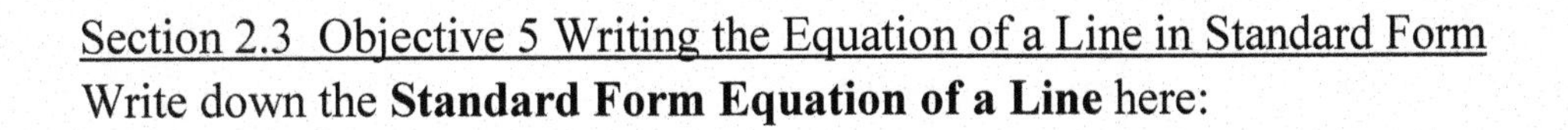

Section 2.3 Objective 5 Writing the Equation of a Line in Standard Form

Write down the **Standard Form Equation of a Line** here:

Write down the **Tip** seen on page 2.3-16:

Work through the **Guided Visualization** titled "Equations of Lines: Standard Form" seen on the bottom of page 2.3-16. Draw a line below with $A = 1$, $B = 3$, and $C = -6$. According the **Guided Visualization,** what is the equation of that line?

Work through the video that accompanies Example 5 and take notes here:
Find the equation of the line passing through the points $(-1,3)$ and $(2,-4)$.
Write the equation in point-slope form, slope-intercept form, and standard form.

Make sure that you know how to write an equation of a line in point-slope form, slope-intercept form, and standard form! Write the point-slope, slope-intercept, and standard forms here:

Point-Slope Form:

Slope-Intercept Form:

Standard Form:

Section 2.3 Objective 6 Finding the Slope and y-Intercept of a Line in Standard Form

Watch the video that accompanies Objective 6 and take notes here:

Given a line of the form $Ax + By = C$, $B \neq 0$, what is the slope of this line and what is the y-intercept?

Slope = ______________ and y-intercept = ____________

Work through the video that accompanies Example 6 and take notes here:

Find the slope and y-intercept and sketch the line $3x - 2y = 6$.

Section 2.3 Objective 7 Sketching Lines by Plotting Intercepts

Watch the video that accompanies Example 7 and take notes here:

Sketch the line $2x - 5y = 8$ by plotting intercepts.

What is the definition of an ***x*-intercept**?

What is the definition of a ***y*-intercept**?

Section 2.3 Objective 8 Finding the Equations of Horizontal and Vertical Lines

Horizontal Lines: Watch the video that describes the equation of a horizontal line and take notes here:

What is the slope of every horizontal line?

What is the equation of a horizontal line?

Vertical Lines: Watch the video that describes the equation of a vertical line and take notes here:

Does a vertical line have slope?

What is the equation of a vertical line?

Watch the video that accompanies Example 8 and take notes here:

a. Find the equation of the horizontal line passing through the point $(-1,3)$.

b. Find the equation of the vertical line passing through the point $(-1,3)$.

Before going on to Section 2.4, you may want to write all of the different types of equations of lines for future reference. These forms are summarized at the end of Section 2.3 in your eText.

Point-Slope Form

Slope-Intercept Form

Standard Form

Horizontal Line

Vertical Line

Section 2.4 Parallel and Perpendicular Lines

- ▯ Work through Objective 1
- ▯ Work through Objective 2
- ▯ Work through Objective 3
- ▯ Work through Objective 4
- ▯ Work through Objective 5

Section 2.4 Parallel and Perpendicular Lines

Section 2.4 Objective 1 Understanding the Definition of Parallel Lines

Write down the Theorem found in Objective 1:

Work through the video that accompanies Example 1 and write your notes here:

Show that the lines $y = -\frac{2}{3}x - 1$ and $4x + 6y = 12$ are parallel.

Section 2.4 Objective 2 Understanding the Definition of Perpendicular Lines

Write down the Theorem found in Objective 2:

Draw and label Figure 30 here:

Work through the video that accompanies Example 2 and write your notes here:

Show that the lines $3x-6y=-12$ and $2x+y=4$ are perpendicular.

Write down the Summary of Parallel and Perpendicular Lines following Example 2 here:

Write down the **Tip** seen on page 2.4-9.

Section 2.4 Objective 3 Determining Whether Two Lines Are Parallel, Perpendicular, or Neither

Watch the video that accompanies Example 3 and take notes here:

For each of the following pairs of lines, determine whether the lines are parallel, perpendicular, or neither.

a. $\begin{array}{l} 3x-y=4 \\ x+3y=7 \end{array}$

b. $\begin{array}{l} y=\frac{1}{2}x+3 \\ x+2y=1 \end{array}$

c. $\begin{array}{l} x=-1 \\ x=3 \end{array}$

Section 2.4 Objective 4 Finding the Equations of Parallel and Perpendicular Lines

You may want to turn back to your notes from Section 2.3 and write down the following equations of lines:

Point-Slope Form

Slope-Intercept Form

Standard Form

Horizontal Line

Vertical Line

Watch the video that accompanies Example 4 and take notes here:

Find the equation of the line parallel to the line $2x+4y=1$ that passes through the point $(3,-5)$. Write the answer in point-slope form, slope-intercept form, and standard form.

Watch the video that accompanies Example 5 and take notes here:
Find the equation of the line perpendicular to the line $y = -5x + 2$ that passes through the point $(3, -1)$. Write the answer in slope-intercept form.

Section 3.1 Guided Notebook

Section 3.1 Relations and Functions

- ☐ Work through Section 3.1 TTK #3
- ☐ Work through Section 3.1 TTK #4
- ☐ Work through Objective 1
- ☐ Work through Objective 2
- ☐ Work through Objective 3
- ☐ Work through Objective 4
- ☐ Work through Objective 5
- ☐ Work through Objective 6

Section 3.1 Relations and Functions

Section 3.1 Objective 1 Understanding the Definitions of Relations and Functions

Watch the video that accompanies the definition of a relation and take notes here:

Write down the definition of a **relation.**

Write an example of a relation as seen in the video and state the domain and range of the relation.

Write down the definition of a **function.**

Read through the text preceding Example 1 to further clarify functions and relations and take notes here.

Work through the video that accompanies Example 1 and take notes here:
Determine whether each relation is a function, and then find the domain and range.

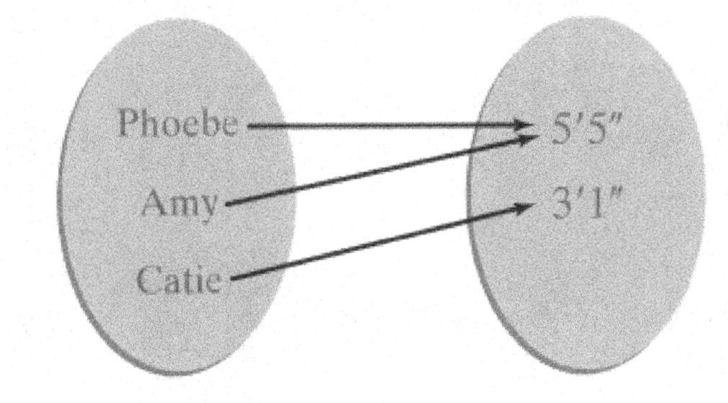

b.

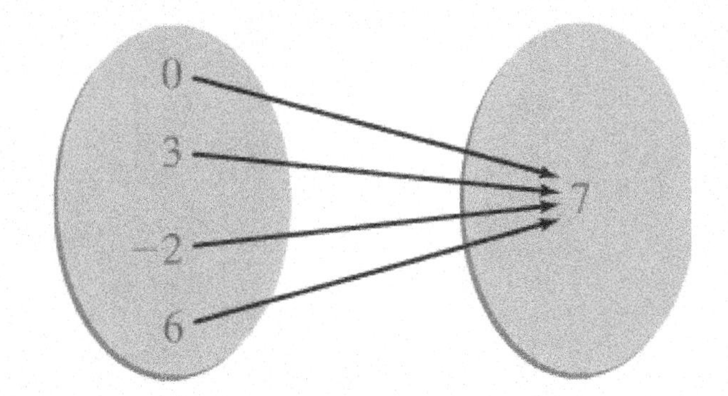

c. $\{(3, 7), (-3, 2), (-4, 5), (1, 4), (3, -4)\}$

d. $\{(-1, 2), (0, 2), (1, 5), (2, 3), (3, 2)\}$

Work through Example 2 and take notes here: Use the domain and range of the following relation to determine whether the relation is a function.

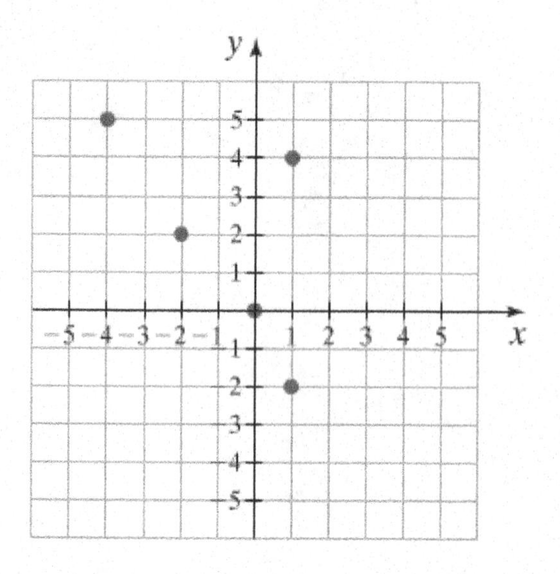

Section 3.1 Objective 2 Determining Whether Equations Represent Functions

Work through the interactive video that accompanies Example 3 and take notes here:

Determine whether the following functions represent y as a function of x.

a. $3x - 2y = -12$

b. $y = 3x^2 - x + 2$

c. $(x + 3)^2 + y^2 = 16$

Section 3.1 Objective 3 Using Function Notation; Evaluating Functions

Carefully take notes from the video that accompanies Objective 3.

What is the definition of an **independent variable**?

What is the definition of a **dependent variable**?

Fill out the following table as seen in this video. (Add more values if you wish.)

Domain Value x	**Range Value** $f(x)=\frac{3}{2}x+6$	**Ordered Pair** $(x, f(x))$

Now sketch the function $f(x)=\frac{3}{2}x+6$ using your values from the table .

Work through the interactive video that accompanies Example 4 and take notes here:

Rewrite these equations using function notation where y is a function of x. Then answer the question following each equation.

a. $3x - y = 5$

What is the value of $f(4)$?

b. $x^2 - 2y + 1 = 0$

Does the point $(-2,1)$ lie on the graph of this function?

c. $y + 7 = 0$

What is $f(x)$ when $x = 3$?

Work through Example 5:

Given that $f(x) = x^2 + x - 1$, evaluate the following:

a. $f(0)$

b. $f(-1)$

c. $f(x+h)$

d. $\dfrac{f(x+h) - f(x)}{h}$ (Note that there is a video solution to part d.)

Section 3.1 Objective 4 Using the Vertical Line Test

Watch the video that accompanies Objective 4 and write your notes here:

In your own words, explain how the **vertical line test** works:

Work through Example 6 in your eText and take notes here: Use the vertical line test to determine which of the following graphs represents the graph of a function.

a.

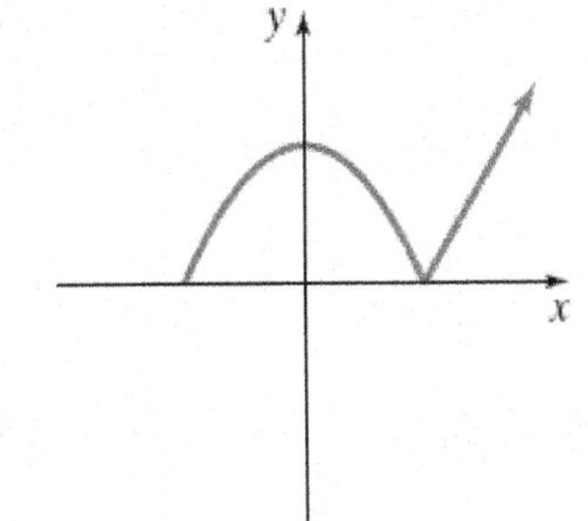

b.

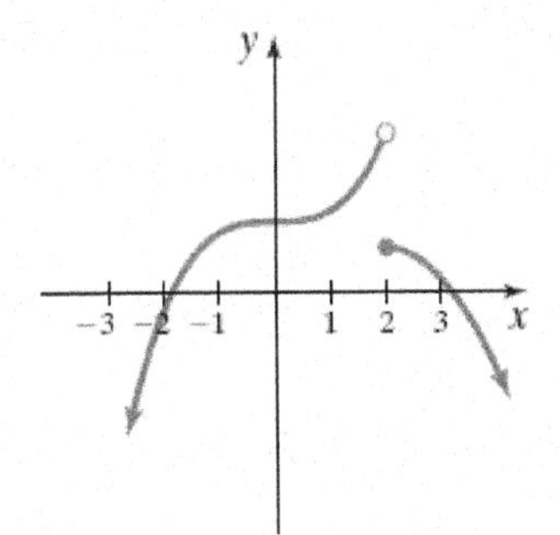

c.

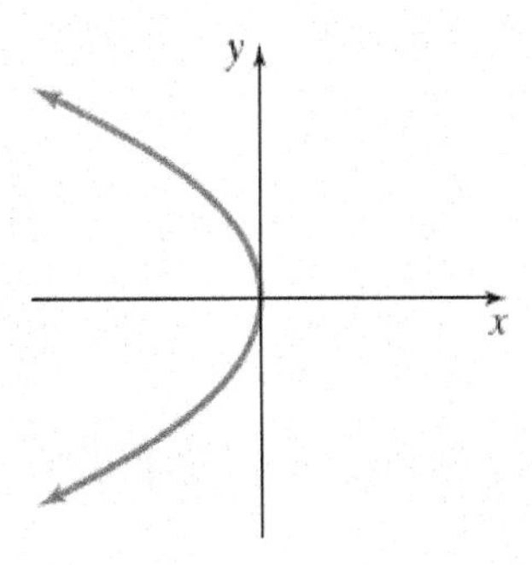

Section 3.1 Objective 5 Classifying Functions

Define a **polynomial function** here:

Give an example of a polynomial function here:

Work through the video that accompanies Example 7 and take notes here:
Determine whether the following functions are polynomial functions. If the function is a polynomial function, state the degree of the polynomial function.

a. $f(x) = 7 - 3x^2 + 5x^3$ b. $h(t) = -11$ c. $f(t) = 3t^6 + \frac{1}{t}$

Define a **rational function** here:

Write down the **Tip** that can be seen below the definition of a rational function.

Give an example of a rational function here:

Work through the video that accompanies Example 8 and take notes here:
Determine whether the following functions are rational functions.

a. $f(t) = \dfrac{t^3 - 1}{t + 1}$

b. $h(x) = (x^3 - x^2 - 9)x^{-5}$

Define a **root function** here:

Give an example of two different root functions. One root function should be an odd root function and the other should be an even root function.

Section 3.1 Objective 6 Determining the Domain of a Function Given the Equation

Carefully work through the video that accompanies Objective 6 and fill in the following blanks:

The **domain of a function** $y = f(x)$ is the set of ____________________ of ___ for which the function is ________________.

The domain of every **polynomial function** is ______________ .

The domain of every **rational function** is the set of all _________ numbers such that ____________ .

The Domain of a Root Function of the Form $f(x) = \sqrt[n]{g(x)}$

1. If n is even, then the domain is the solution to the inequality _______________ .

2. If n is odd, then the domain is the set of all __________ numbers for which __________ is defined.

Below is a summary of the three classifications of functions seen in the previous video and a summary of how to find the domain of each. Use this summary for future reference.

Class of Function	Form	Domain
Polynomial functions	$f(x) = a_n x^n + a_{n-1}x^{n-1} + \cdots + a_1 x + a_0$	Domain is $(-\infty, \infty)$.
Rational functions	$f(x) = \frac{g(x)}{h(x)}$, where $g \neq 0$ and h are polynomial functions such that the degree of $h(x)$ is greater than 0.	Domain is all real numbers such that $h(x) \neq 0$.
Root functions	$f(x) = \sqrt[n]{g(x)}$, where $g(x)$ is a function and n is an integer such that $n \geq 2$.	1. If n is even, the domain is the solution to the inequality $g(x) \geq 0$. 2. If n is odd, the domain is the set of all real numbers for which g is defined.

Work through interactive video that accompanies Example 9 and explain how to find the domain of each of the following functions:

a. $f(x) = 2x^2 - 5x$

b. $f(x) = \dfrac{x}{x^2 - x - 6}$

c. $h(x) = \sqrt{x^2 - 2x - 8}$

d. $f(x) = \sqrt[3]{5x - 9}$

e. $k(x) = \sqrt[4]{\dfrac{x-2}{x^2 + x}}$

Section 3.2 Guided Notebook

Section 3.2 Properties of a Function's Graph

- ☐ Work through Objective 1
- ☐ Work through Objective 2
- ☐ Work through Objective 3
- ☐ Work through Objective 4
- ☐ Work through Objective 5
- ☐ Work through Objective 6

Section 3.2 Properties of a Function's Graph

Section 3.2 Objective 1 Determining the Intercepts of a Function

Work through the video that accompanies Objective 1 and fill in the notes below:

Define ***y*-intercept**:

Define ***x*-intercept**:

Given a function $f(x)$, how can we find the intercepts?

Show how to find the intercepts of the functions $f(x) = -3x + 2$ and $g(x) = x^3 - x^2 - 12x$.

Work through the video that accompanies Example 2 and show your work here:

Given the functions below, find the x-intercept(s) and the y-intercept (if they exist).

a. $f(x)=x^3-2x^2+x-2$

b. $g(x)=\dfrac{2x^2-7x-4}{x^3}$

Section 3.2 Objective 2 Determining the Domain and Range of a Function from Its Graph

Watch the video that accompanies Objective 2 and take notes here:

The examples seen in the video are the **same** graphs used in Example 3:
Find the domain and range of the functions seen below:

a.

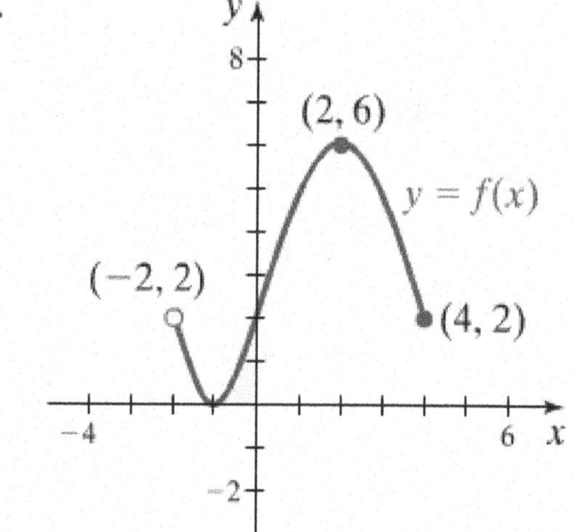

Domain:

Range:

b.

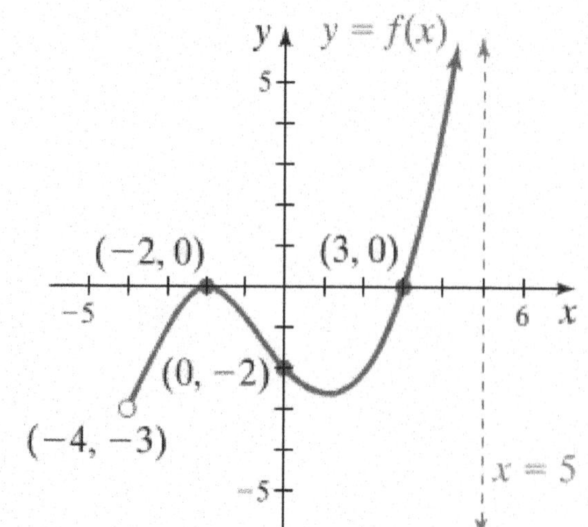

Domain:

Range:

c.

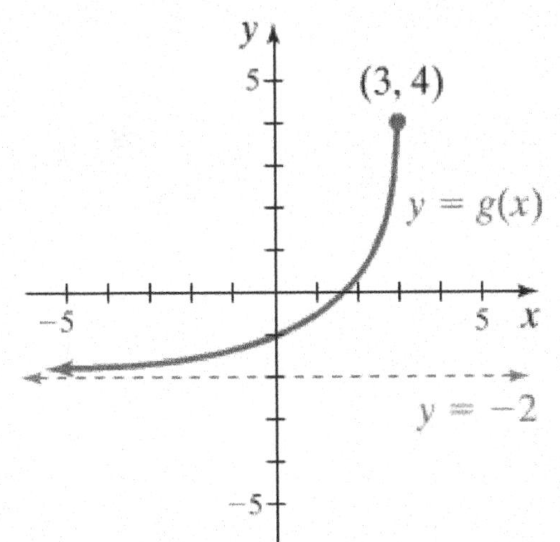

Domain:

Range:

What is the definition of a **vertical asymptote**? (See the link in the solution to Example 3 part b.)

What is the definition of a **horizontal asymptote**? (See the link in the solution to Example 3 part c.)

Section 3.2 Objective 3 Determining Whether a Function is Increasing, Decreasing, or Constant

Watch the video that accompanies Objective 3 and fill in the following definitions:

Increasing:

Decreasing:

Constant

Watch the video that accompanies Example 4 and determine the intervals on which the function seen below is increasing, decreasing, and constant: (**The interval(s) are always *x*-intervals.**)

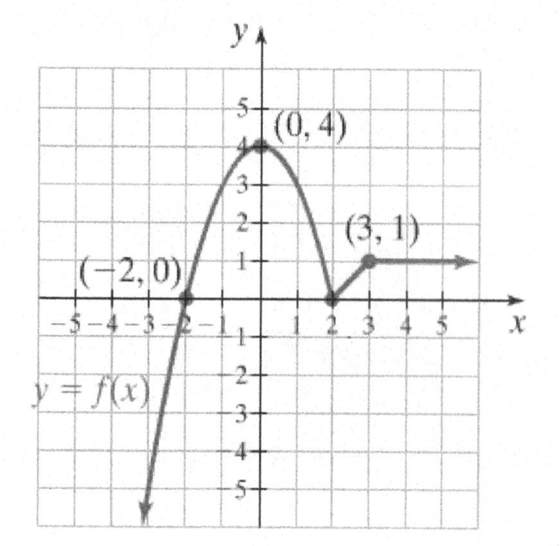

The function is increasing on the interval(s):

The function is decreasing on the interval(s):

The function is constant on the interval(s):

Section 3.2 Objective 4 Determining Relative Maximum and Relative Minimum Values of a Function

Watch the video that accompanies Objective 4 and fill in the following definitions:

Relative Maximum:

Relative Minimum:

Work through the video that accompanies Example 5 and answer each of the following questions:

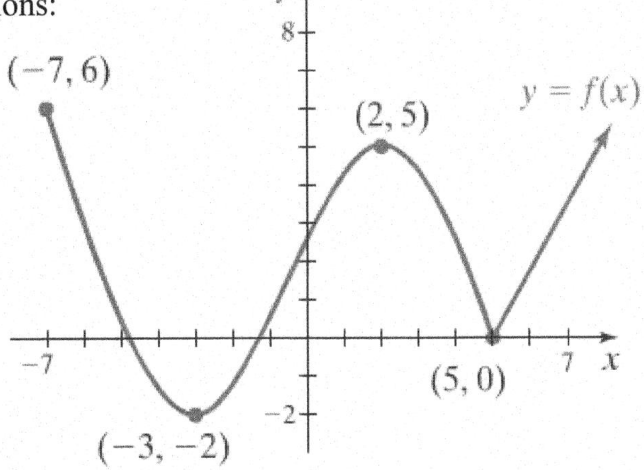

a. On what interval(s) is f increasing?

b. On what interval(s) is f decreasing?

c. For what value(s) of x does f have a relative minimum?

d. For what value(s) of x does f have a relative maximum?

e. What are the relative minima?

f. What are the relative maxima?

Section 3.2 Objective 5 Determining Whether a Function is Even, Odd, or Neither

Watch the video that accompanies Objective 5 and fill in the following definitions:

Even Functions:

(Draw an example of an even function here.)

Odd Functions:

(Draw an example of an odd function here.)

Summarize Even and Odd Functions by filling in the following Table.

TYPE OF FUNCTION	ALGEBRAIC DEFINITION	TYPE OF SYMMETRY (Write y-axis or origin symmetry.)
Even	$f(-x) =$ ____________	
Odd	$f(-x) =$ ____________	

Work through Example 6:

Determine whether each function is even, odd, or neither. (Explain in your own words why each function is even, odd, or neither.)

a.

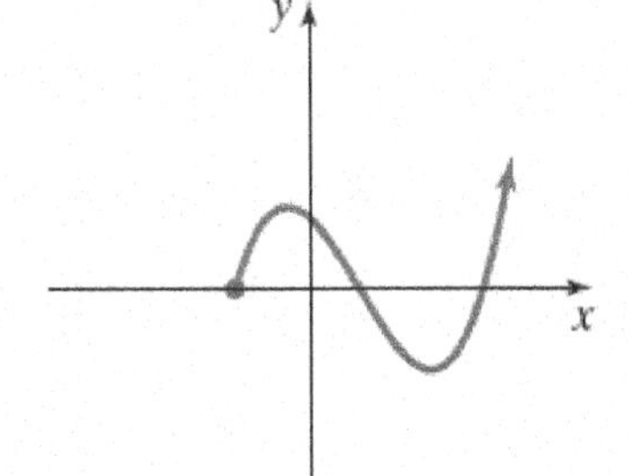

b.

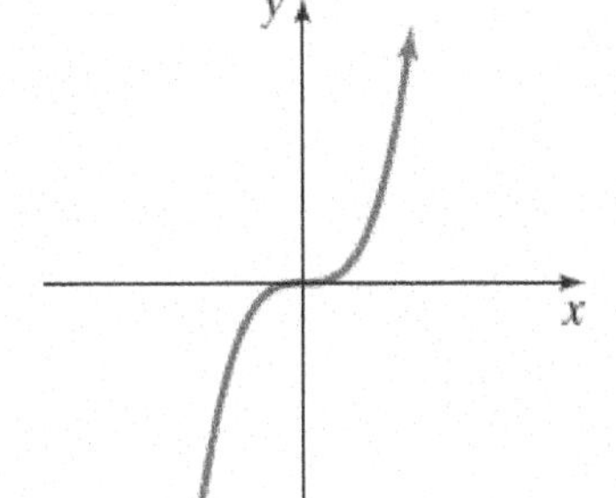

c.

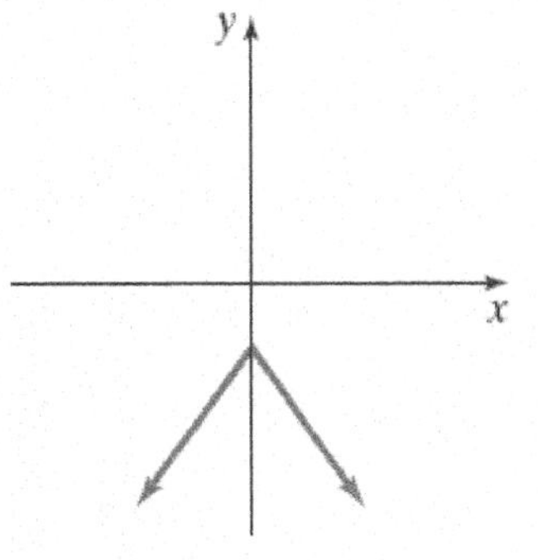

You do not need the graph of a function to determine if it is even, odd, or neither!

Carefully, work through the video that accompanies Example 7 and determine (*without graphing*) whether each of the following functions is even, odd, or neither:

a. $f(x)=x^3+x$

b. $g(x)=\frac{1}{x^2}+7|x|$

c. $h(x)=2x^5-\frac{1}{x}$

d. $G(x)=x^2+4x$

<u>Section 3.2 Objective 6 Determining Information about a Function from a Graph</u>

Work through the animation that accompanies Example 8 and answer each question:

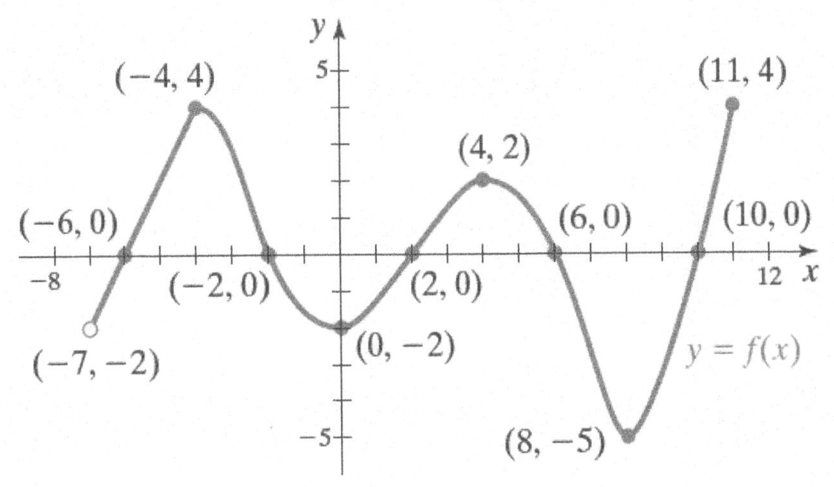

a. What is the y-intercept?

b. What are the real zeros of f?

c. Determine the domain and range of f.

d. Determine the interval(s) on which f is increasing, decreasing and constant.

e. For what value(s) of x does f obtain a relative maximum? What are the relative maxima?

f. For what value(s) of x does f obtain a relative minimum? What are the relative minima?

g. Is f even, odd, or neither?

h. For what values of x is $f(x) = 4$?

i. For what values of x is $f(x) < 0$?

Section 3.3 Guided Notebook

Section 3.3 Graphs of Basic Functions; Piecewise Functions

- ☐ Work through Objective 1
- ☐ Work through Objective 2
- ☐ Work through Objective 3
- ☐ Work through Objective 4

Section 3.3 Graphs of Basic Functions; Piecewise Functions

Section 3.3 Objective 1 Sketching the Graphs of the Basic Functions

Read through all the text in Objective 1. **YOU MUST MEMORIZE THE GRAPHS** of the 9 basic functions seen on these pages! On the next three pages of this notebook, sketch each of these functions and list the properties of each. Click on the appropriate link in your eText to see the properties of each function.

1. The constant function $f(x)=b$

2. The identity function $f(x)=x$

3. The square function $f(x)=x^2$

4. The cube function $f(x)=x^3$

5. The absolute value function $f(x)=|x|$

6. The square root function $f(x) = \sqrt{x}$

7. The cube root function $f(x) = \sqrt[3]{x}$

8. The reciprocal function $f(x) = \dfrac{1}{x}$

9. The greatest integer function $f(x) = [\![x]\!]$

Section 3.3 Objective 2 Sketching the Graphs of Basic Functions With Restricted Domains

Sketch the graph of $f(x) = x^2$.

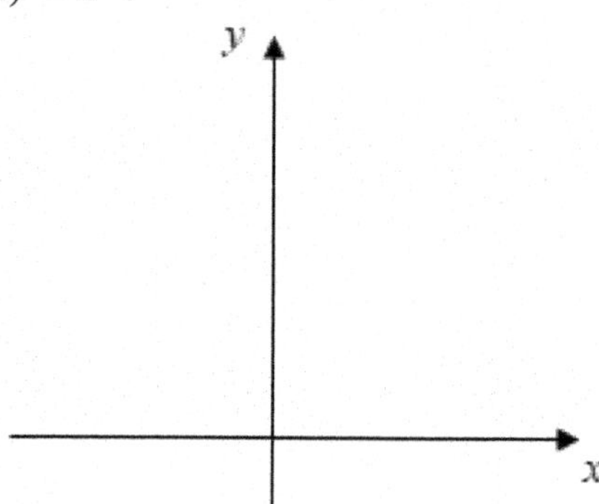

Now sketch the graph of $f(x) = x^2$ for $x > 0$.

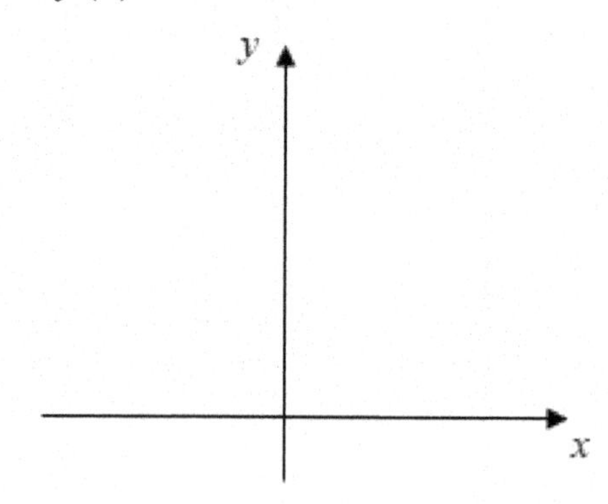

Fill in the blank:

The function $f(x) = x^2$ for $x > 0$ is known as a function with a ____________________ domain.

Work through the video that accompanies Example 1 and show your work here:

Sketch the function $f(x)=\sqrt{x}$ for $x \geq 1$.

Work through the video that accompanies Example 2 and show your work here:

The graph of a basic function with a restricted domain is given below.

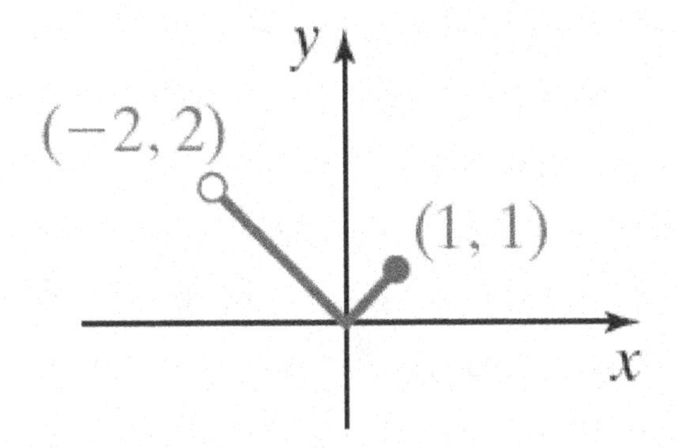

a. Write the domain of the function using interval notation.

b. Define the function using an inequality to express the restricted domain.

Section 3.3 Objective 2 Analyzing Piecewise-Defined Functions

Sketch the graph of the restricted function $y = -x$ for $x < 0$.

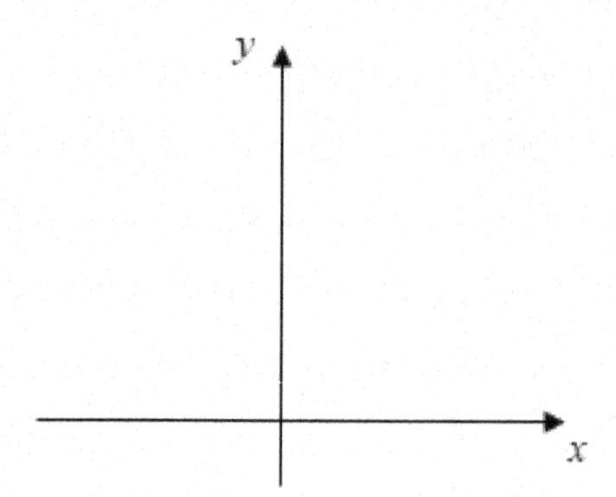

Now, sketch the graph of the restricted function $y = x$ for $x \geq 0$.

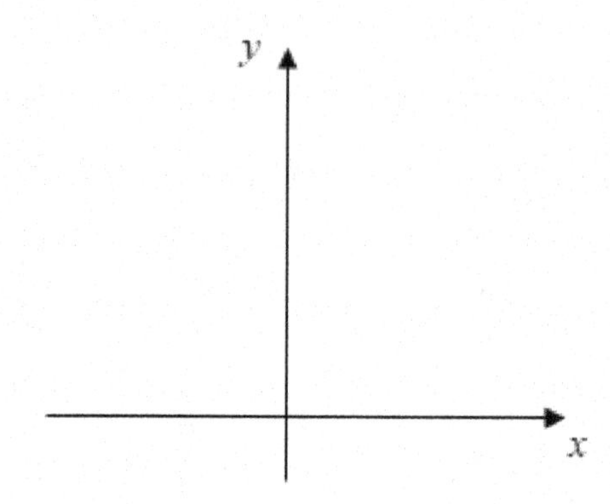

Now, draw **both** of the above two graphs onto the grid below:

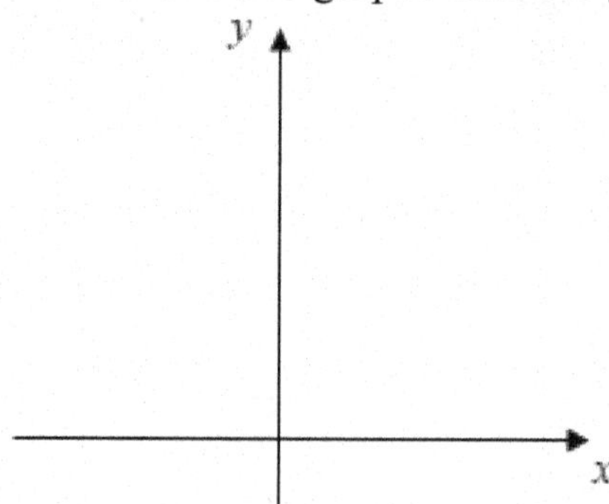

This is the graph of the absolute value function. Define the absolute function by filling in the blanks below:

$$f(x) = |x| = \begin{cases} y = -x \text{ for } \_\_\_\_\_\_\_\_\_\_ \\ y = \ x \text{ for } \_\_\_\_\_\_\_\_\_\_ \end{cases}.$$

What is the definition of a **piecewise-defined function?**

Carefully work through the **animation** that accompanies Example 3 and take notes here: (You should use a pencil to sketch piecewise functions. Why? Watch the animation to find out.)

Sketch the function $f(x)=\begin{cases} x^2 & \text{if } x<1 \\ 1-x & \text{if } x\geq 1 \end{cases}$

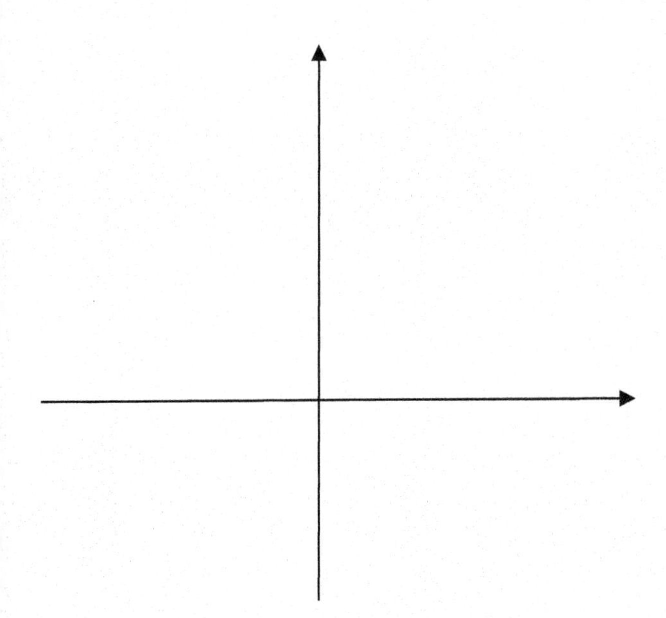

Work through the video that accompanies Example 4 and take notes here:

$$\text{Let } f(x)=\begin{cases} 1 & \text{if } x<-1 \\ \sqrt[3]{x} & \text{if } -1\le x<0 \\ \dfrac{1}{x} & \text{if } x>0 \end{cases}$$

a. Evaluate $f(-3)$, $f(-1)$, and $f(2)$.

b. Sketch the graph of f.

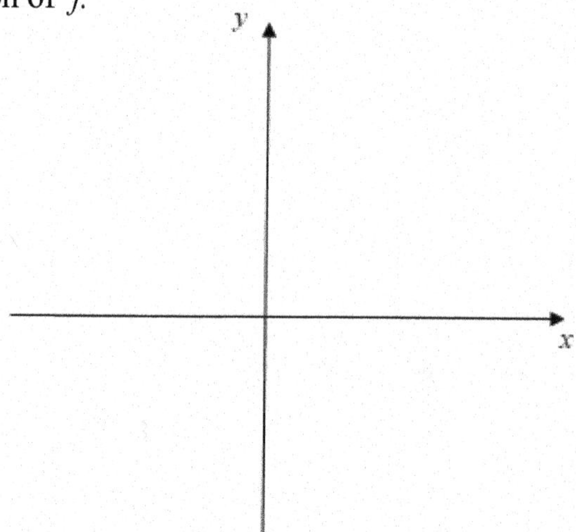

c. Determine the domain of f.

d. Determine the range of f.

Work through the video that accompanies Example 5 and take notes here:
Find a rule that describes the following piecewise function:

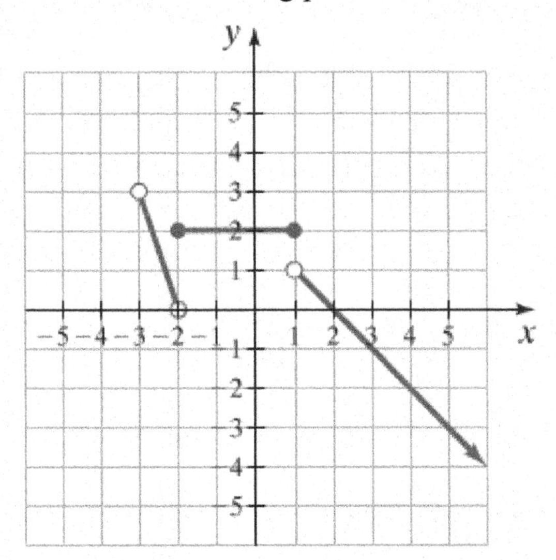

Section 3.3 Objective 4 Solving Applications of Piecewise-Defined Functions

Work through Example 6 taking notes here:

Ted Cruz, a presidential candidate during the 2016 Republican presidential primaries, proposed a flat tax to replace the existing U.S. income tax system. In his tax proposal, every adult would pay $0.00 in taxes on the first $36,000 earned. They would then pay a flat tax of 10% on everything over $36,000. Cruz's tax plan can be described using a piecewise-defined function.

a. According to Cruz's plan, how much in taxes are owed for someone earning $62,500?

b. Find the piecewise function, $T(x)$, that describes the amount of taxes paid, T, as a function of the dollars earned, x, for Cruz's tax plan.

c. Sketch the piecewise function, $T(x)$.

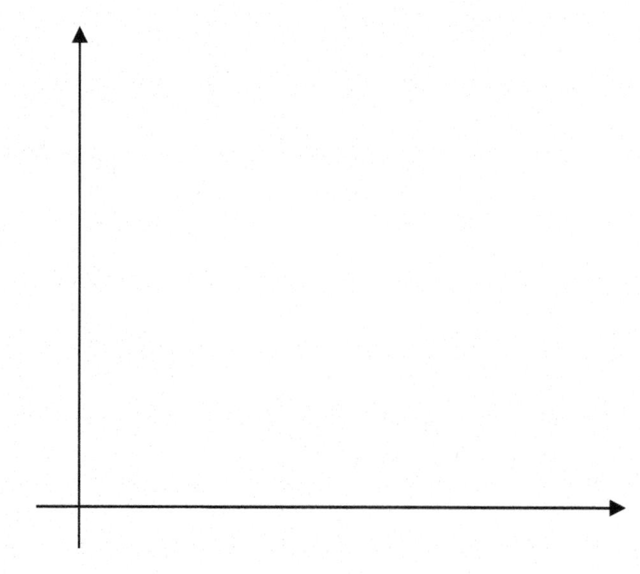

Work through the video that accompanies Example 7 taking notes here:

Cheapo Rental Car Co. charges a flat rate of $85 to rent a car for up to 2 days. The company charges an additional $20 for each additional day (or part of a day).

a. How much does it cost to rent a car for 2 ½ days? 3 days?

b. Write and sketch the piecewise function, $C(x)$, that describes the cost of renting a car as a function of time, x, in days rented for values of x less than or equal to 5.

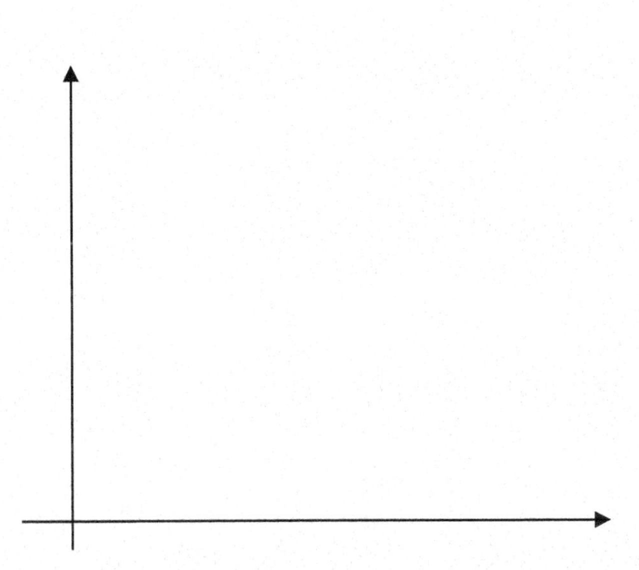

Section 3.4 Guided Notebook

Section 3.4 Transformations of Functions

- ☐ Work through Section 3.4 TTK #1
- ☐ Work through Section 3.4 TTK #4
- ☐ Work through Objective 1
- ☐ Work through Objective 2
- ☐ Work through Objective 3
- ☐ Work through Objective 4
- ☐ Work through Objective 5
- ☐ Work through Objective 6
- ☐ Work through Objective 7

Section 3.4 Transformations of Functions

Section 3.4 Objective 1 Using Vertical Shifts to Graph Functions

Work through the video that accompanies Example 1 and take notes here:
Sketch the graphs of $f(x)=|x|$ and $g(x)=|x|+2$.

If *c* > *0,* explain in your own words how to sketch the graph of $y=f(x)+c$ and $y=f(x)-c$. Click on the "**animate buttons**" on p. 3.4-6 and use the given graph of $y=f(x)$ below to sketch the graphs of $y=f(x)+c$ and $y=f(x)-c$.

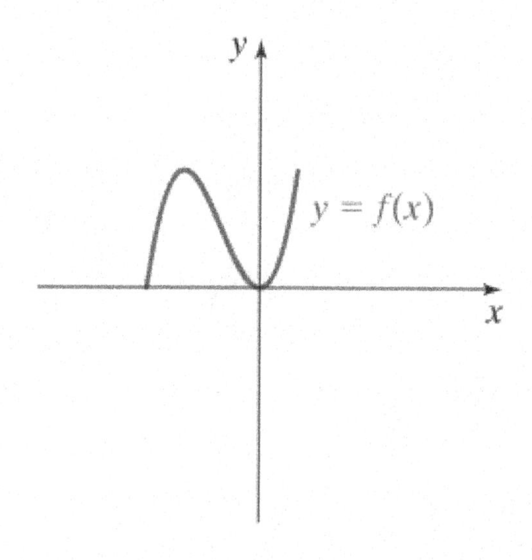

Work through the **Guided Visualization** titled “Vertical Transformations of Functions” seen on page 3.4-7. Then, sketch each of the following functions below:

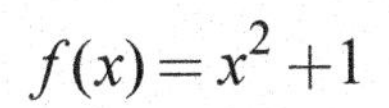

$f(x)=x^2+1$

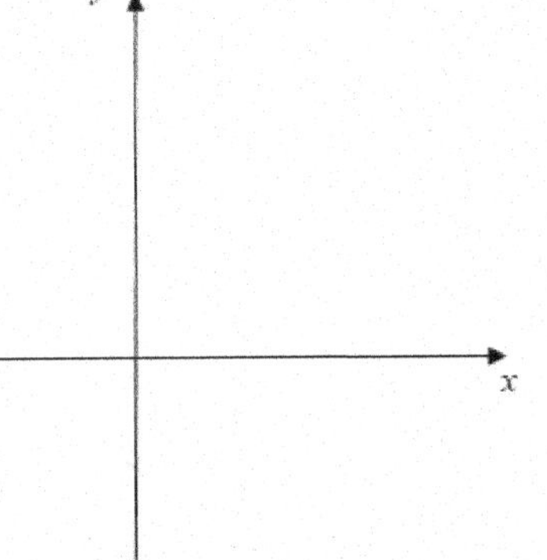

$f(x)=x^3-2$

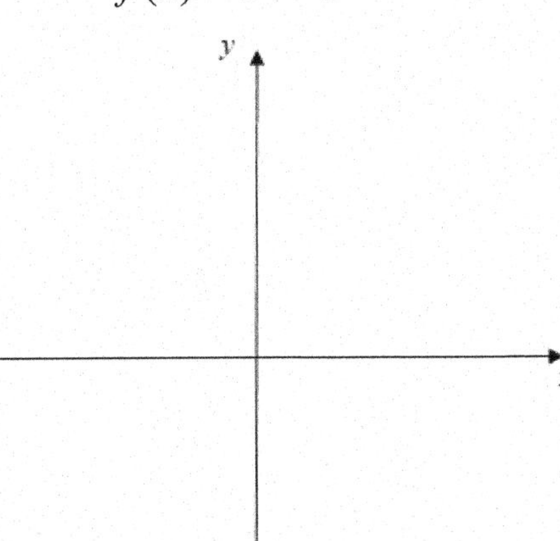

$f(x)=|x|-3$

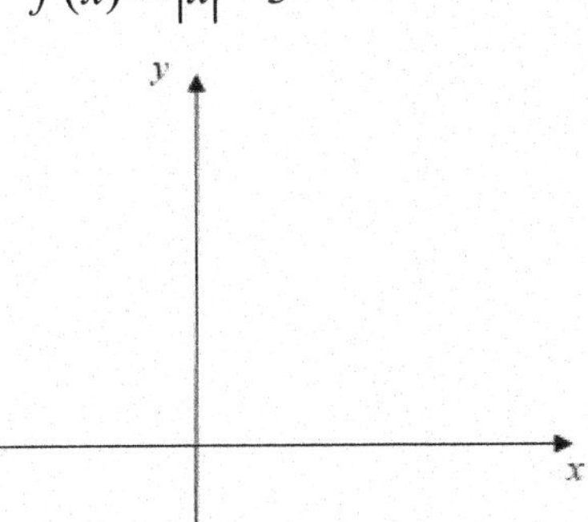

$f(x)=\sqrt{x}+4$

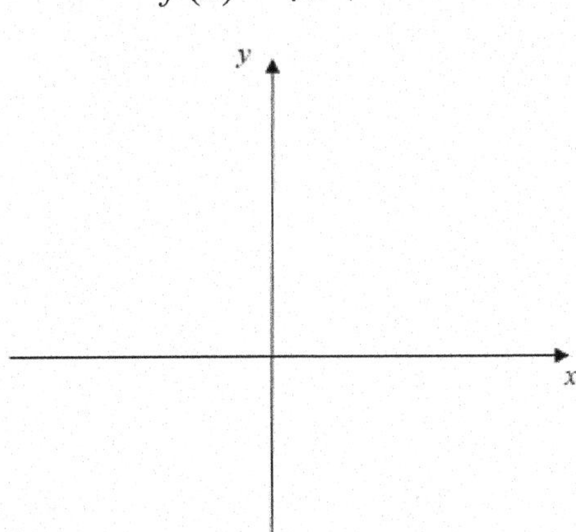

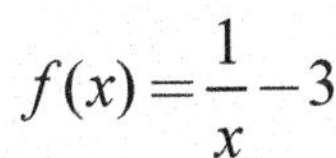

$f(x)=\frac{1}{x}-3$

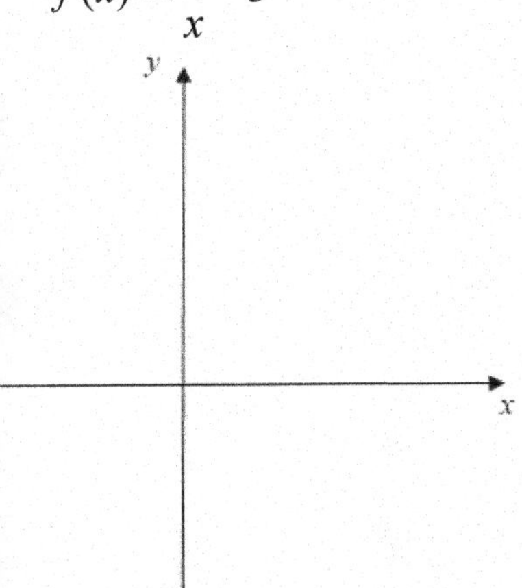

Section 3.4 Objective 2 Using Horizontal Shifts to Graph Functions

Work through the video that accompanies Objective 2 and take notes here:

If $c > 0$, explain in your own words how to sketch the graph of $y = f(x+c)$ and $y = f(x-c)$.

Click on the "**animate buttons**" on p. 3.4-9 and use the given graph of $y = f(x)$ below to sketch the graphs of $y = f(x+c)$ and $y = f(x-c)$.

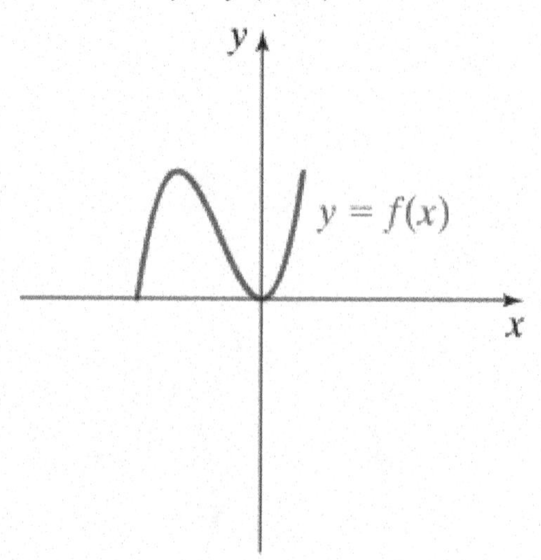

Work through the **Guided Visualization** titled "Horizontal Transformations of Functions" seen on page 3.4-10. Then, sketch each of the following functions below:

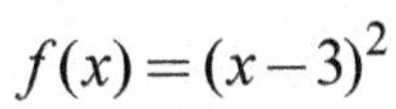

$f(x) = (x-3)^2$

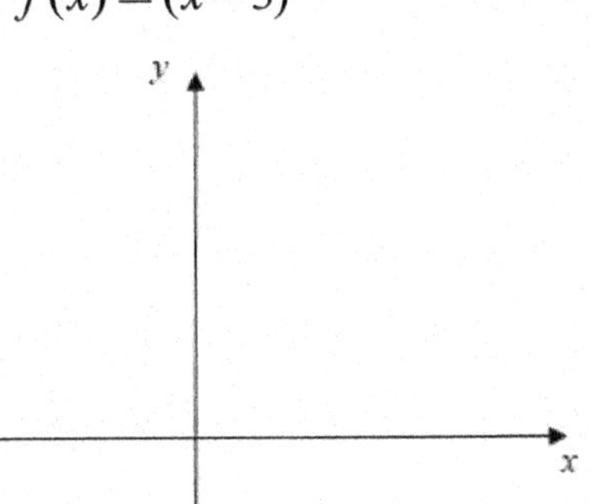

$f(x) = (x+4)^3$

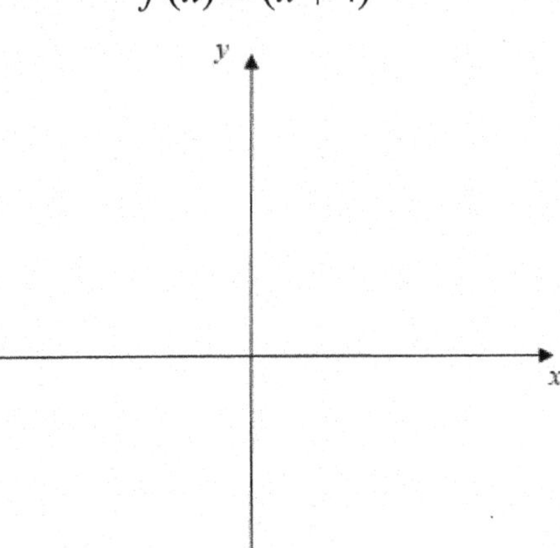

$f(x) = |x+2|$

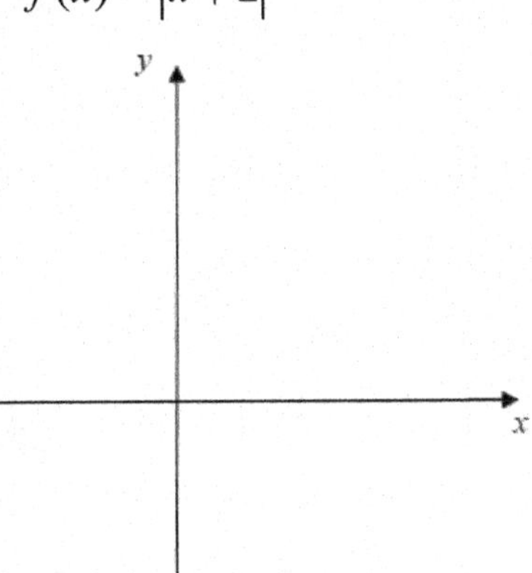

$f(x) = \sqrt{x-3}$

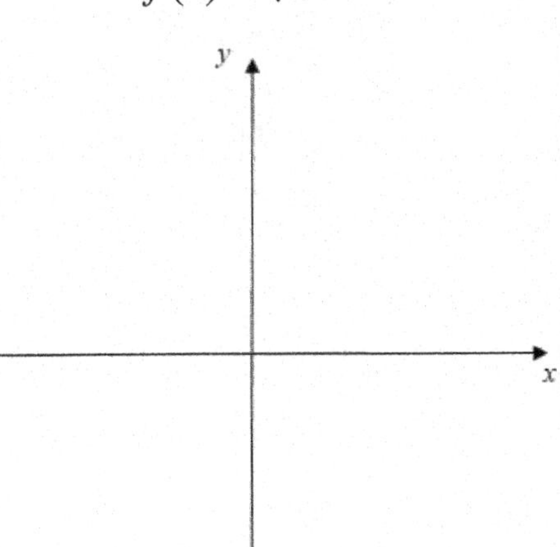

$f(x) = \dfrac{1}{x+1}$

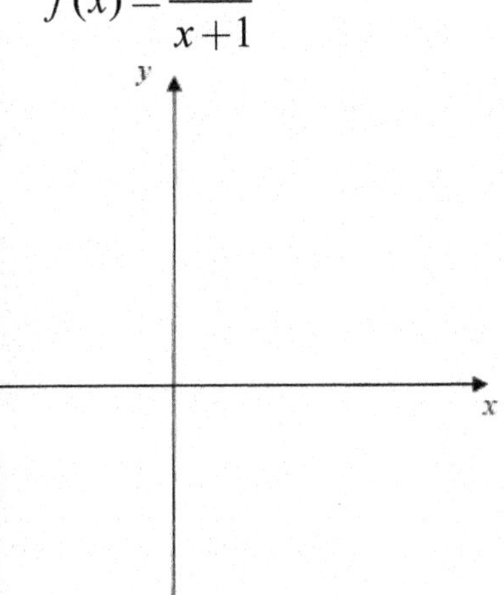

IT IS NOW TIME TO COMBINE A HORIZONTAL AND A VERTICAL SHIFT.

Work through the animation that accompanies Example 2 and take notes here:
Use the graph of $y = x^3$ to sketch the graph of $g(x) = (x-1)^3 + 2$.

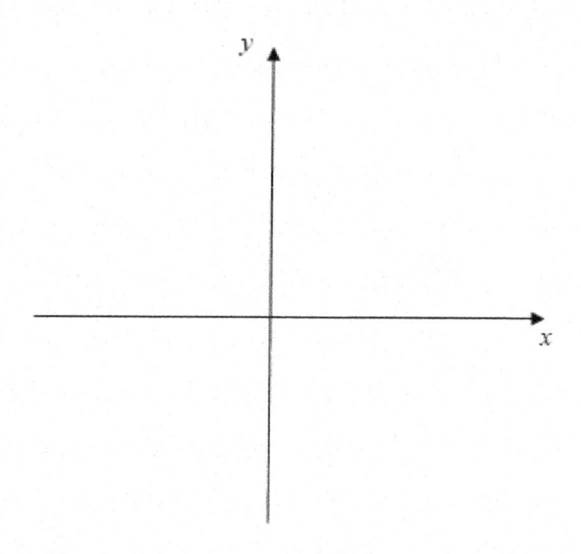

Section 3.4 Objective 3 Using Reflections to Graph Functions

Work through the video that accompanies Objective 3 and take notes here:

If $c > 0$, explain in your own words how to sketch the graph of $y = -f(x)$ and $y = f(-x)$.

Click on the "**animate buttons**" on p. 3.4-13 and p. 3.4-15 and use the given graphs of $y = f(x)$ below to sketch the graphs of $y = -f(x)$ and $y = f(-x)$.

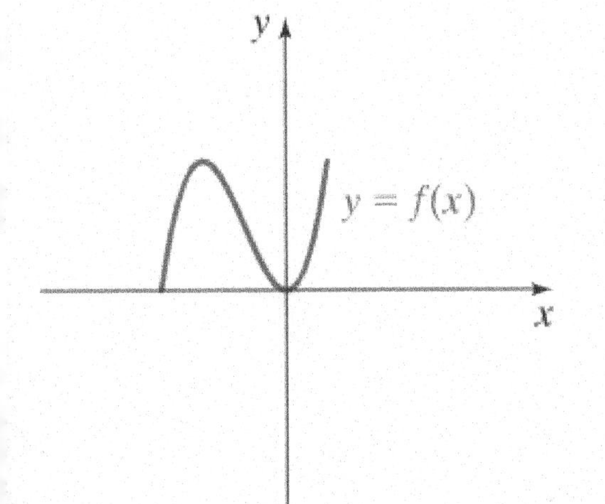

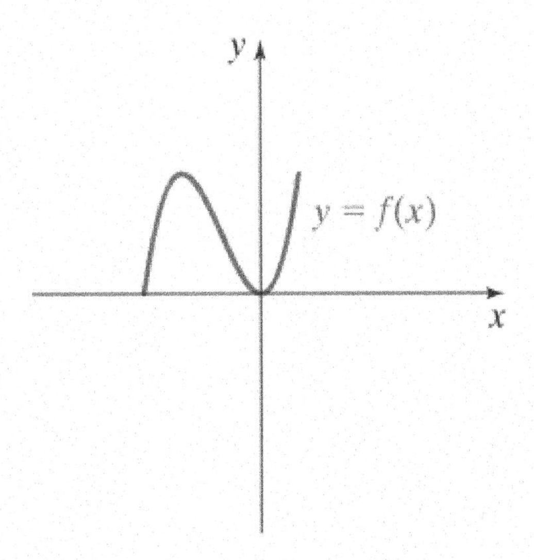

Work through Example 3 and take notes here: Use the graph of the basic function $y=\sqrt[3]{x}$ to sketch each graph.

a) $y=-\sqrt[3]{x}-2$

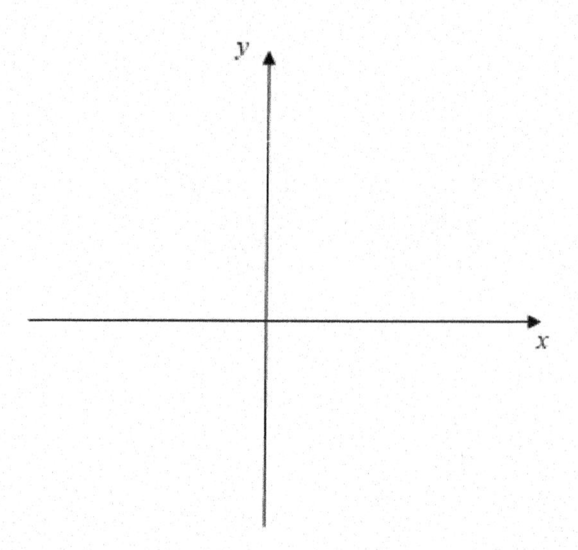

b) $y=\sqrt[3]{1-x}$

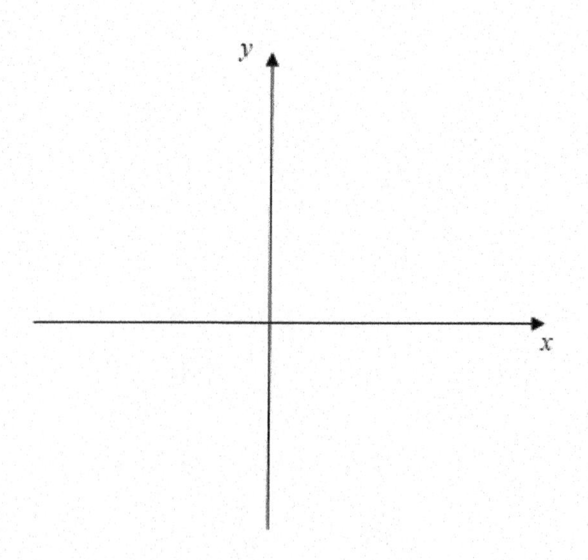

Section 3.4 Objective 4 Using Vertical Stretches and Compressions to Graph Functions

Work through the video that accompanies Example 4 and take notes here:

Use the graph of $f(x)=x^2$ to sketch the graph of $g(x)=2x^2$.

If $a>1$, explain in your own words how to sketch the graph of $y=af(x)$.

If $0<a<1$, explain in your own words how to sketch the graph of $y=af(x)$.

Click on the "**animate buttons**" on p. 3.4-20 and use the given graph of $y = f(x)$ below to sketch the graphs of $y = a\,f(x)$ for $a > 1$ and $y = a\,f(x)$ for $0 < a < 1$.

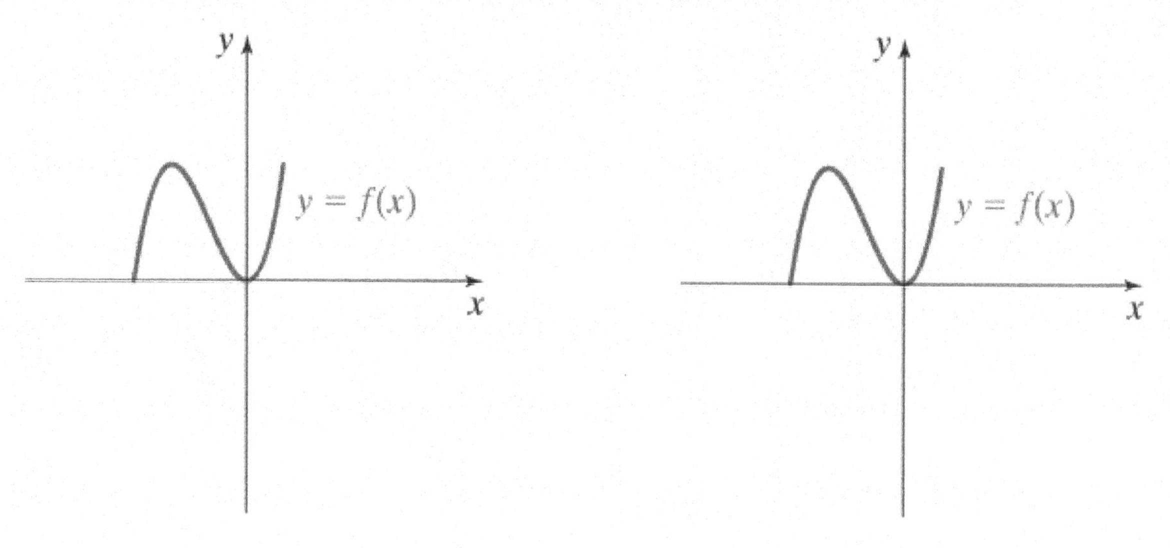

Section 3.4 Objective 5 Using Horizontal Stretches and Compressions to Graph Functions

Work through the video that accompanies Objective 5 and take notes here:

If $a > 1$, explain in your own words how to sketch the graph of $y = f(ax)$.

If $0 < a < 1$, explain in your own words how to sketch the graph of $y = f(ax)$.

Click on the "**animate buttons**" on p. 3.4-22 and use the given graph of $y = f(x)$ below to sketch the graphs of $y = f(ax)$ for $a > 1$ and $y = f(ax)$ for $0 < a < 1$.

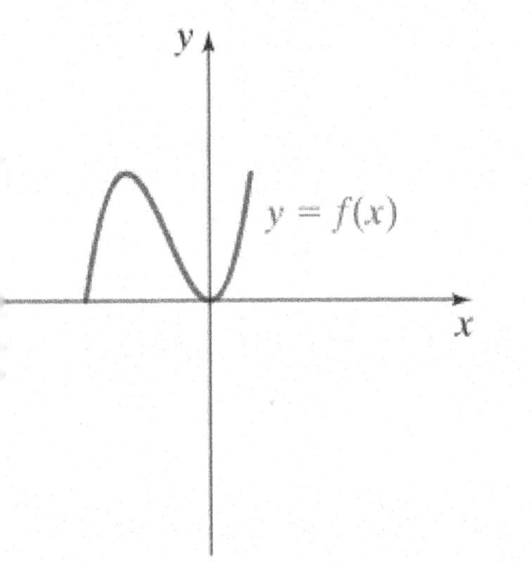

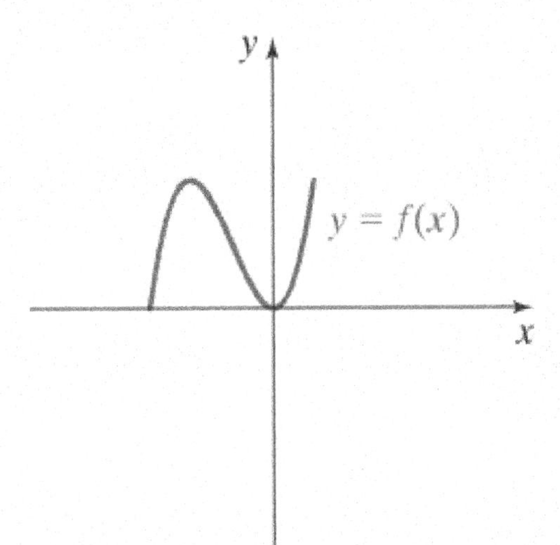

Work through the video that accompanies Example 5 and take notes here:

Use the graph of $f(x)=\sqrt{x}$ to sketch the graphs of $g(x)=\sqrt{4x}$ and $h(x)=\sqrt{\frac{1}{4}x}$.

<u>Section 3.4 Objective 6 Using Combinations of Transformations to Graph Functions</u>

You have learned 6 transformations in this section. You may encounter functions that combine many (if not all) of these transformations. Write down the "order of operations" of transformations as in Objective 6.

"Order of Operations" for Transformations

1.

2.

3.

4.

5.

6.

Work through the animation that accompanies Example 6 and take notes here:

Use transformations to sketch the graph of $y = -2(x+3)^2 - 1$.

Work through the interactive video that accompanies Example 7 and take notes here:
Use the graph of $y = f(x)$ to sketch each of the following functions.

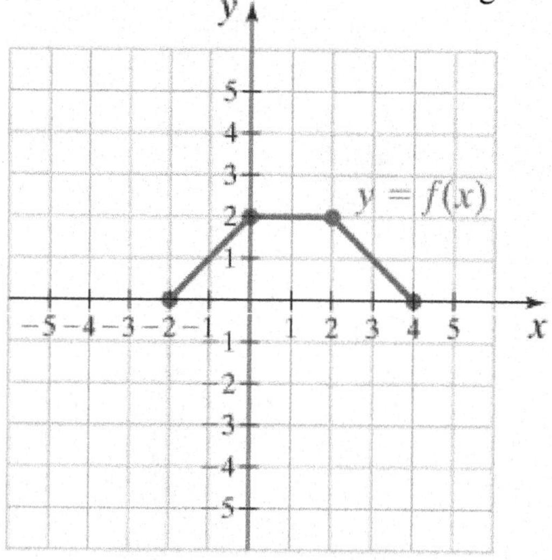

a) $y = -f(2x)$

b) $y = 2f(x-3)-1$

c) $y = -\frac{1}{2}f(2-x)+3$

Section 3.4 Objective 7 Using Transformations to Sketch the Graphs of Piecewise-Defined Functions

Work through the video that accompanies Example 8 and take notes here:

Sketch the graph of the function $f(x)=\begin{cases} -|x+2| & \text{if } x<-1 \\ \sqrt{x+1} & \text{if } x\geq -1 \end{cases}$.

Use the **Summary of Transformation Techniques** to complete the following statements:

Given a function $y = f(x)$ and a constant $c > 0$:

1. The graph of $y = f(x) + c$ is obtained by...

2. The graph of $y = f(x) - c$ is obtained by...

3. The graph of $y = f(x + c)$ is obtained by...

4. The graph of $y = f(x - c)$ is obtained by...

5. The graph of $y = -f(x)$ is obtained by...

6. The graph of $y = f(-x)$ is obtained by...

7. Suppose a is a positive real number. The graph of $y = af(x)$ is obtained by...

8. Suppose a is a positive real number. The graph of $y = f(ax)$ is obtained by...

Section 3.5 Guided Notebook

Section 3.5 The Algebra of Functions; Composite Functions

- ☐ Work through Objective 1
- ☐ Work through Objective 2
- ☐ Work through Objective 3
- ☐ Work through Objective 4
- ☐ Work through Objective 5

Section 3.5 The Algebra of Functions; Composite Functions

Work through the video in the introduction and learn about the Algebra of Functions.

Write down 4 definitions below and give one example of each:

1. The sum of f and g

2. The difference of f and g

3. The product of f and g

4. The quotient of f and g

Section 3.5 Objective 1 Evaluating a Combined Function

Work through Example 1 and take notes here:

Let $f(x)=\dfrac{12}{2x+4}$ and $g(x)=\sqrt{x}$.

Find each of the following:

a. $(f+g)(1)$

b. $(f-g)(1)$

c. $(fg)(4)$

d. $\left(\dfrac{f}{g}\right)(4)$

Work through the interactive video that accompanies Example 2 and take notes here:
Use the graph to evaluate each expression or state that it is undefined:

a. $(f+g)(1)$

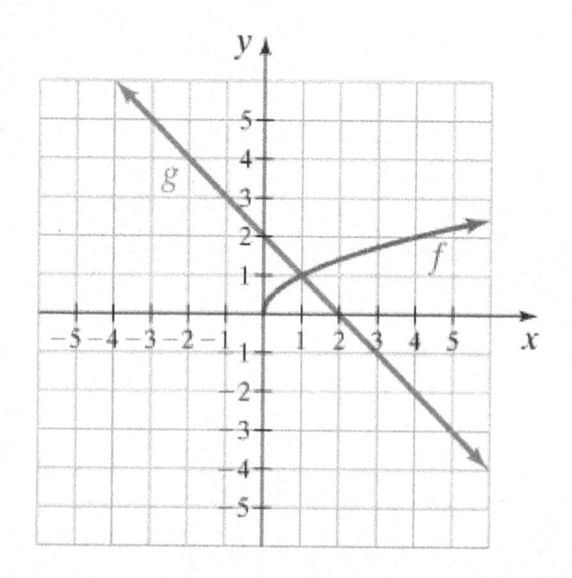

b. $(f-g)(0)$

c. $(fg)(4)$

d. $\left(\dfrac{f}{g}\right)(2)$

<u>Section 3.5 Objective 2 Finding the Intersection of Intervals</u>

In order to find the domain of a combined function, you *must* be able to find the intersection of two or more intervals. Watch the video that accompanies Example 3 and take notes here:

Find the intersection of the following sets and graph the set on a number line.

a) $[0,\infty)\cap(-\infty,5]$

b) $((-\infty,-2)\cup(-2,\infty))\cap[-4,\infty)$

Section 3.5 Objective 3 Finding Combined Functions and Their Domains

Work through the video that accompanies Objective 3 and take notes here:

Work through the video that accompanies Example 4 and take notes here:
Let $f(x)=\sqrt{x+1}$ and $g(x)=x^2-16$.

a. Find $f+g$ and determine the domain of $f+g$.

b. Find $\frac{f}{g}$ and determine the domain of $\frac{f}{g}$.

Work through the video that accompanies Example 5 and take notes here:

Let $f(x)=\dfrac{x+2}{x-3}$ and $g(x)=\sqrt{4-x}$.

a. Find $f+g$ and determine the domain of $f+g$.

b. Find $f-g$ and determine the domain of $f-g$.

c. Find fg and determine the domain of fg.

d. Find $\dfrac{f}{g}$ and determine the domain of $\dfrac{f}{g}$.

Section 3.5 Objective 4 Forming and Evaluating Composite Functions

Carefully work through the video that accompanies Objective 4 and take notes here. **Make sure that you can define the composite function:**

Work through the interactive video that accompanies Example 6 and take notes here:

Let $f(x) = 4x + 1$, $g(x) = \frac{x}{x-2}$, and $h(x) = \sqrt{x+3}$.

a) Find the function $f \circ g$.

b) Find the function $g \circ h$.

c) Find the function $h \circ f \circ g$.

d) Evaluate $(f \circ g)(4)$, or state that it is undefined.

e) Evaluate $(g \circ h)(1)$, or state that it is undefined.

f) Evaluate $(h \circ f \circ g)(6)$, or state that it is undefined.

Work through the interactive video that accompanies Example 7 and take notes here:
Use the graph to evaluate each expression or state that it is undefined:

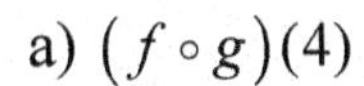

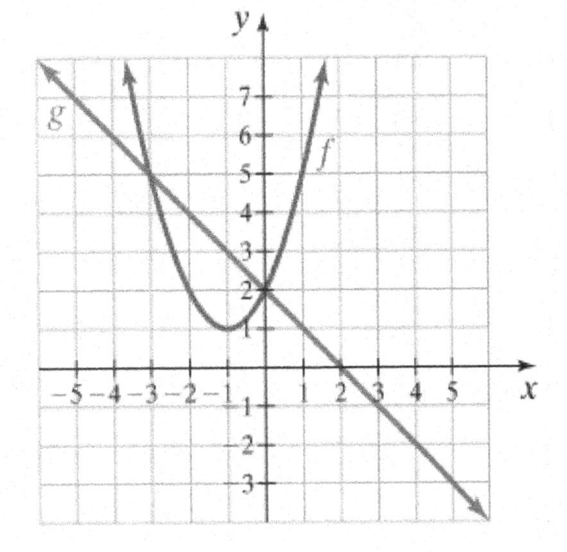

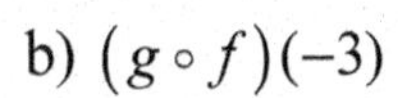

c) $(f \circ f)(-1)$

d) $(g \circ g)(4)$

e) $(f \circ g \circ f)(1)$

Section 3.5 Objective 5 Determining the Domain of Composite Functions

Work through the video that accompanies Objective 5 and take notes here. Carefully write down how to find the domain of a composite function and give an example of how to find the domain of a composite function.

Work through the interactive video that accompanies Example 8 and take notes here.

Let $f(x)=\dfrac{-10}{x-4}$, $g(x)=\sqrt{5-x}$, and $h(x)=\dfrac{x-3}{x+7}$.

a) Find the domain of $f\circ g$.

b) Find the domain of $g\circ f$.

c) Find the domain of $f\circ h$.

d) Find the domain of $h\circ f$.

Section 3.6 Guided Notebook

Section 3.6 One-to-One Functions; Inverse Functions

- ▯ Work through Objective 1
- ▯ Work through Objective 2
- ▯ Work through Objective 3
- ▯ Work through Objective 4
- ▯ Work through Objective 5

Section 3.6 One-to-One Functions; Inverse Functions

Introduction to Section 3.6

In your own words, what does the function $F(C)=\frac{9}{5}C+32$ represent?

In your own words, what does the function $C(F)=\frac{5}{9}(F-32)$ represent?

If $F(C)=\frac{9}{5}C+32$, then evaluate $F(100)$. What does mean $F(100)$?

If $C(F)=\frac{5}{9}(F-32)$, then evaluate $C(212)$. What does mean $C(212)$?

What is the value of $C(F(100))$?

What is the value of $F(C(212))$?

Section 3.6 Objective 1 Understanding the Definition of a One-to-One Function

Work through the video that accompanies Objective 1 and write notes here.

Write down the definition of a **one-to-one function:**

Give an example of a function that is one-to-one.

Give an example of a function that is **not** one-to-one.

Write down the **Alternate Definition of a One-to-One Function** as seen in the eText.

Section 3.6 Objective 2 Determining If a Function is One-to-One Using the Horizontal Line Test

Work through the video that accompanies Objective 2 and write notes here.

Write down the **Horizontal Line Test** and write down the 3 examples seen in the video.

Work through the animation that accompanies Example 1 and take notes here:
Determine whether each function is one-to-one.

a.

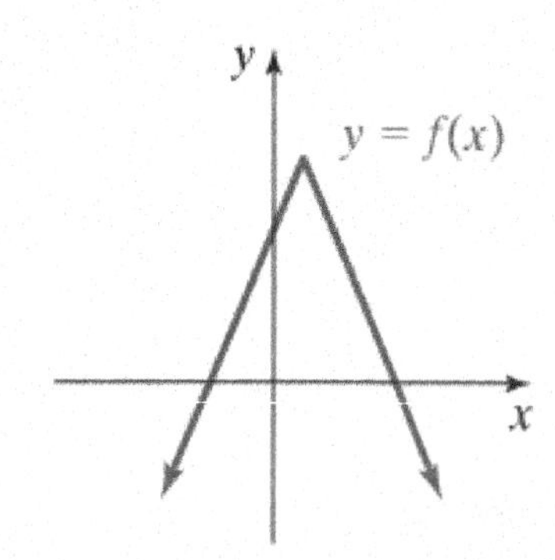

b.

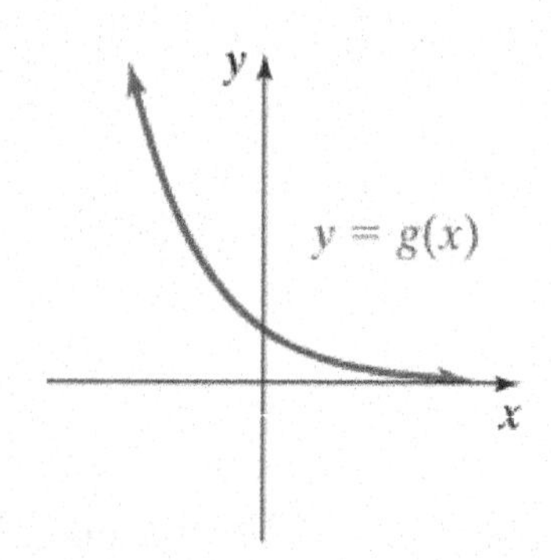

c. $f(x)=x^2+1,\ x\leq 0$ (You must sketch the graph of this function.)

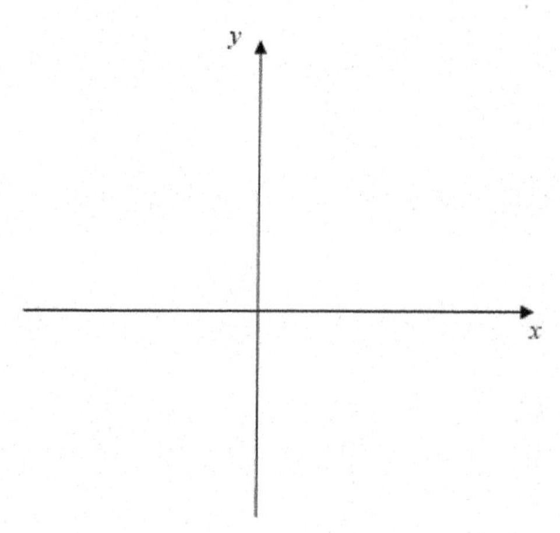

d. $f(x)=\begin{cases}2x+4 & \text{for } x\leq -1\\ 2x-6 & \text{for } x\geq 4\end{cases}$ (You must sketch the graph of this function.)

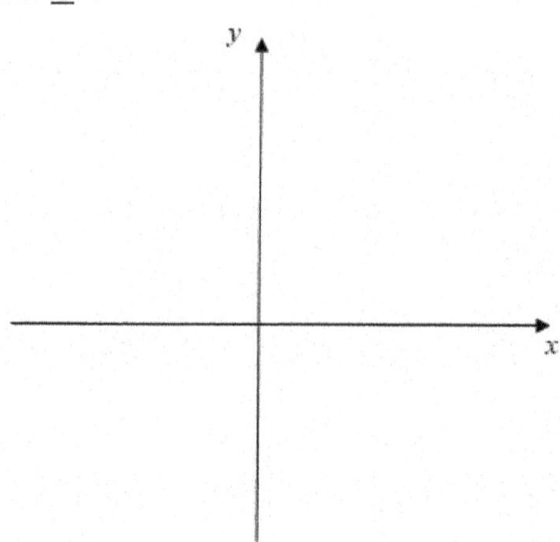

Section 3.6 Objective 3 Understanding and Verifying Inverse Functions

Work through the video that accompanies Objective 3 and write notes here.

Write down the definition of an **inverse function.**

Write down the example of the one-to-one function and the inverse function given in this video.

Watch the video that follows Figure 27 on page 3.6-10 and take notes here:

Write down the two **Composition Cancellation Equations**:

Work through the interactive video that accompanies Example 2 and take notes here:

Show that $f(x)=\dfrac{x}{2x+3}$ and $g(x)=\dfrac{3x}{1-2x}$ are inverse functions using the composition cancellation equations.

Section 3.6 Objective 4 Sketching the Graphs of Inverse Functions

Read through Objective 4 and describe in your own words how to sketch the graph of the inverse of a given one-to-one function.

Below are the graphs of two one-to-one functions. Sketch the graphs of their inverse functions.

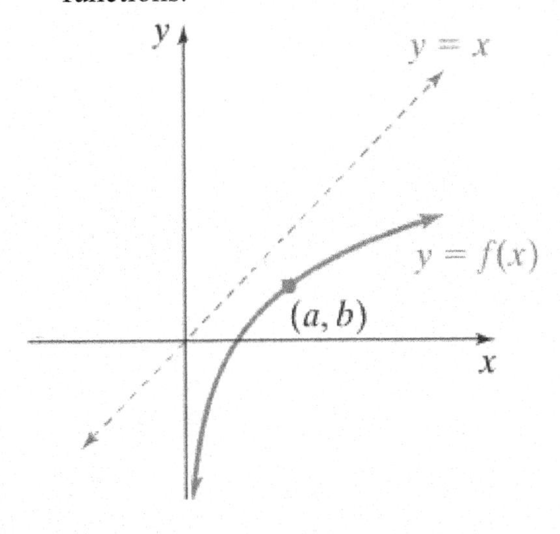

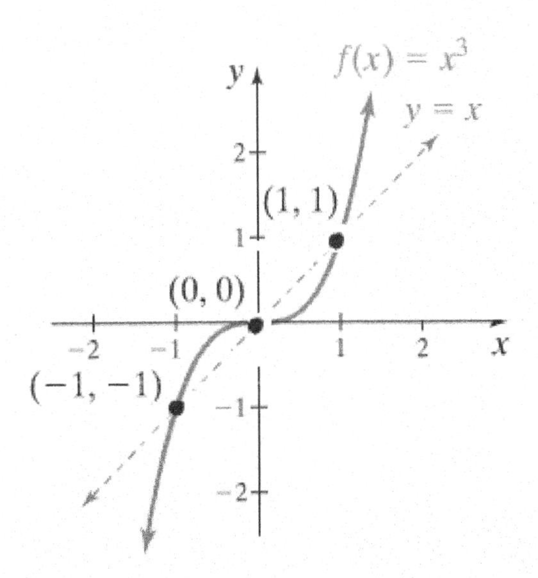

Work through the animation that accompanies Example 3:
Sketch the graph of $f(x) = x^2 + 1,\ x \le 0$, and its inverse.
Also state the domain and range of f and f^{-1}.

Click on the **Guided Visualization** titled "Sketching the Graphs of Inverse Functions" found on page 3.6-15. For each graph below, sketch the graph of $y = x$, sketch and label the graph of the inverse function, include at least one ordered pair that lies on the graph of f, and include at least one ordered pair that lies on the graph of f^{-1}.

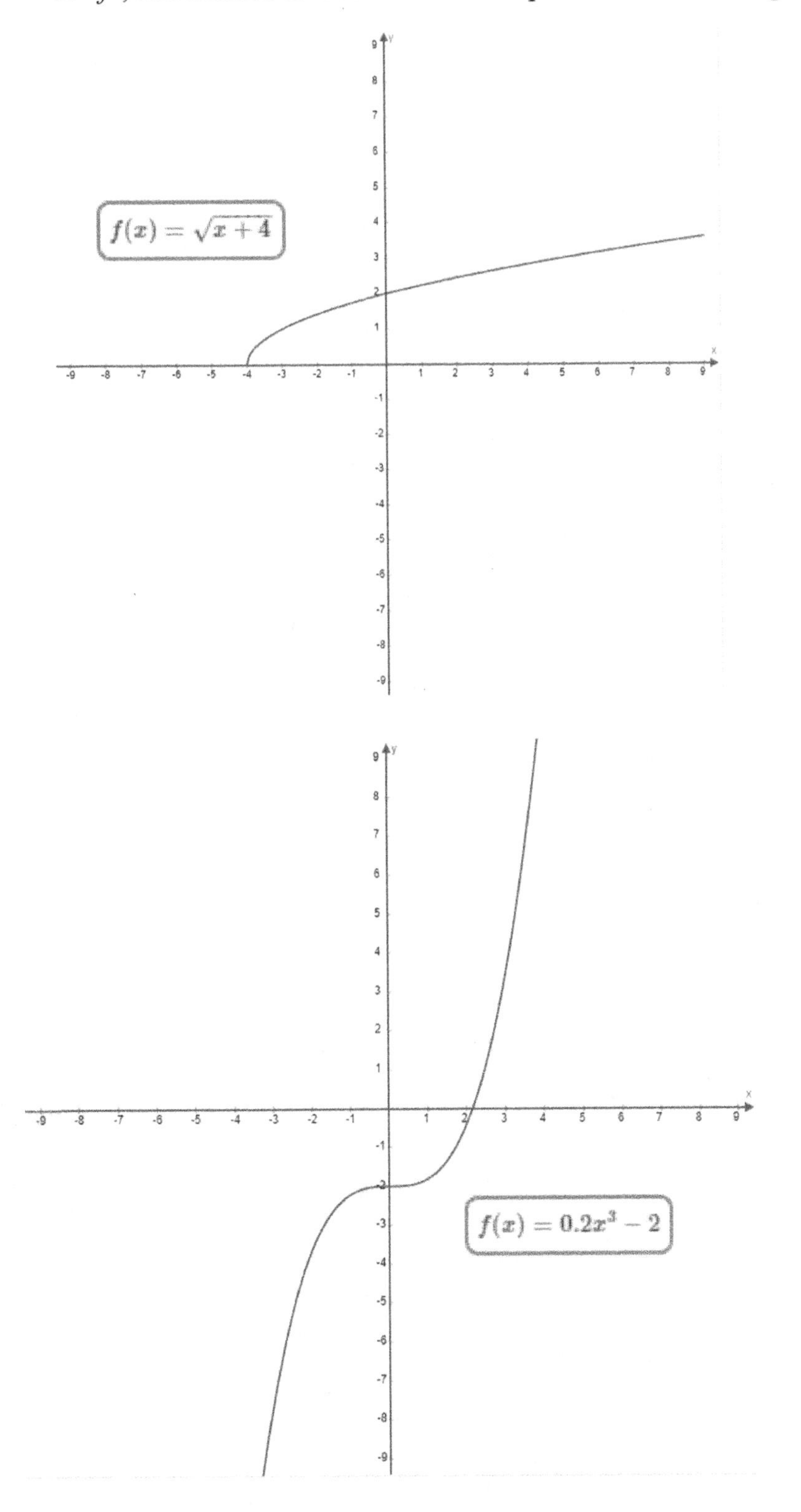

Section 3.6 Objective 5 Finding the Inverse of a One-to-One Function

Work through the animation that accompanies Example 4: Find the inverse function of $f(x)=\dfrac{2x}{1-5x}$ and state the domain and range of f and f^{-1}

Step 1. Change $f(x)$ to y :

Step 2. Interchange x and y :

Step 3: Solve for y:

Step 4: Change y to $f^{-1}(x)$:

Write the domain and range of f and f^{-1}.

Work through the video that accompanies Example 5.
Find the inverse of $f(x) = x^2 + 1,\ x \leq 0$. Write down the four steps for finding inverse functions as you find the inverse of $f(x) = x^2 + 1,\ x \leq 0$

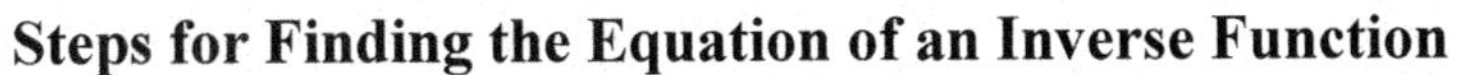

Steps for Finding the Equation of an Inverse Function

Step 1:

Step 2:

Step 3:

Step 4:

In your own words, explain the relationship between the domain and range of a one-to-one function and its inverse function:

Use the **Inverse Function Summary** to complete the following statements:

1. The function f^{-1} exists if and only if…

2. The domain of f is the same as the…

   And the range of f is the same as the…

3. To verify that two one-to-one functions, f and g, are inverses of each other, we must…

4. The graph of f^{-1} is a reflection of …

   That is, for any point (a,b) that lies on the graph of f, the point (b,a) must…

5. To find the inverse of a one-to-one function,…

Section 4.1 Guided Notebook

Section 4.1 Quadratic Functions

- ☐ Work through Section 4.1 TTK #1
- ☐ Work through Section 4.1 TTK #2
- ☐ Work through Section 4.1 TTK #3
- ☐ Work through Section 4.1 TTK #5
- ☐ Work through Objective 1
- ☐ Work through Objective 2
- ☐ Work through Objective 3
- ☐ Work through Objective 4
- ☐ Work through Objective 5

Section 4.1 Quadratic Functions

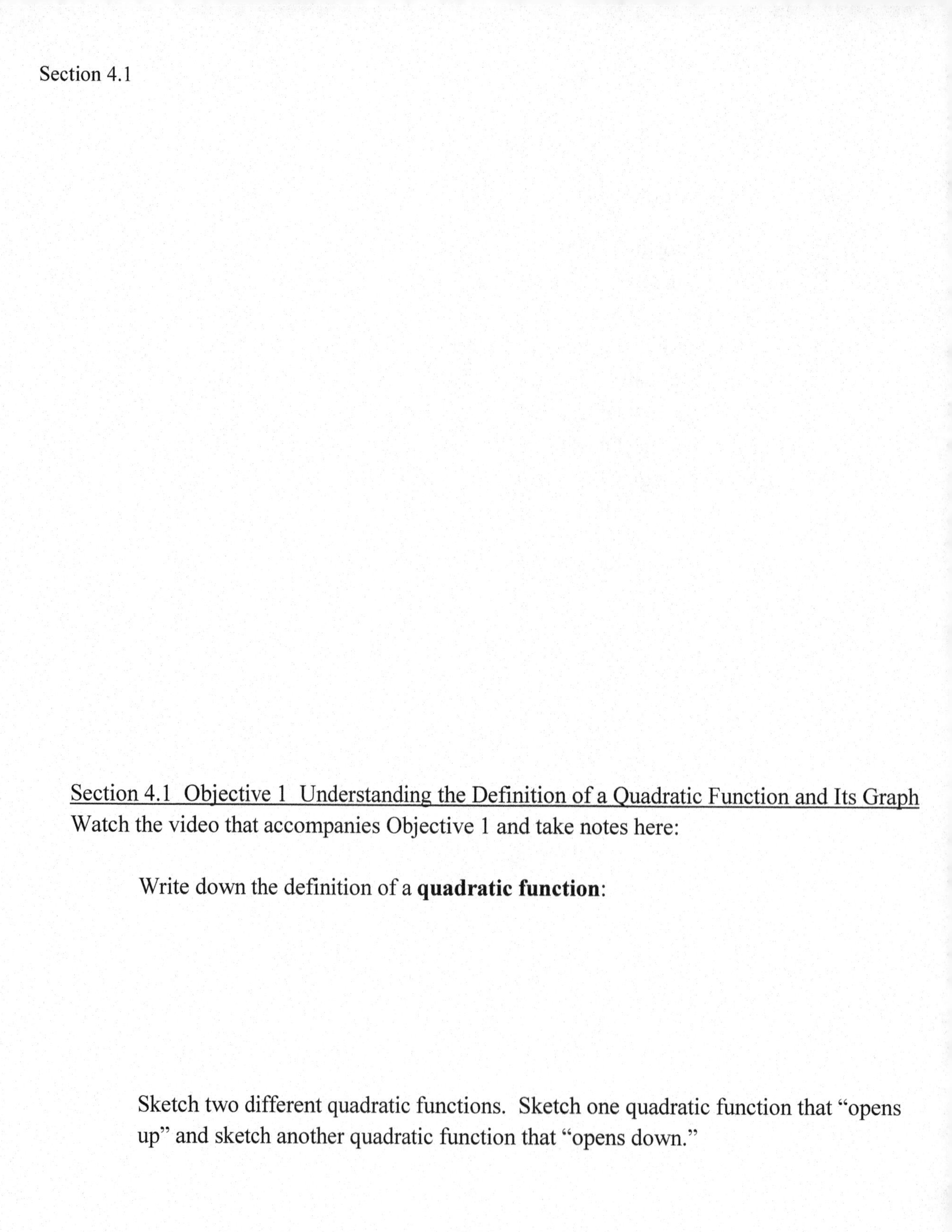

Section 4.1 Objective 1 Understanding the Definition of a Quadratic Function and Its Graph

Watch the video that accompanies Objective 1 and take notes here:

Write down the definition of a **quadratic function**:

Sketch two different quadratic functions. Sketch one quadratic function that "opens up" and sketch another quadratic function that "opens down."

What determines whether or not the graph of a quadratic function of the form $f(x) = ax^2 + bx + c$ opens up or down?

Work through Example 1 and take notes here: Without graphing, determine whether the graph of the quadratic function $f(x) = -3x^2 + 6x + 1$ opens up or down.

It is crucial that you understand the five basic characteristics of a parabola. Carefully work through the animation and describe the following five characteristics of a parabola in your own words.

1. Vertex

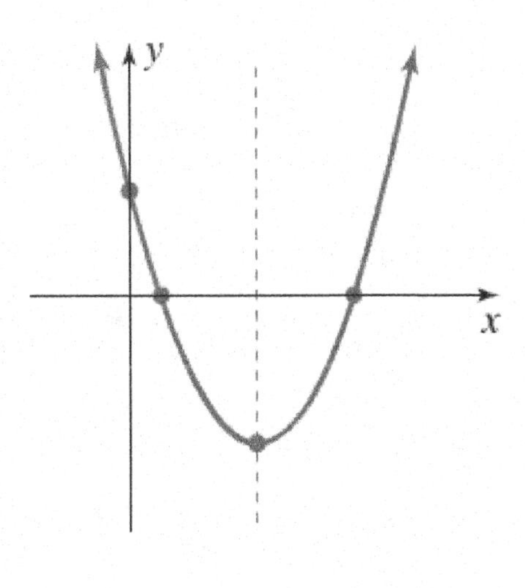

2. Axis of Symmetry

3. y-intercept

4. x-intercept(s) or real zeros

5. Domain and range

Section 4.1 Objective 2 Graphing Quadratic Functions Written in Vertex form

Work through the animation and explain how to sketch the graph of $f(x) = -2(x+3)^2 - 1$.

Vertex Form of a Quadratic Function (Fill in the Blanks)

A quadratic function is in **vertex form** if it is written as

$f(x) =$ __________________________. The graph is a parabola with vertex _________.

The parabola "opens up" if _______. The parabola "opens down" if ________.

Work through the video that accompanies Example 2 and answer each of the following questions:

Given that the quadratic function $f(x) = -(x-2)^2 - 4$ is in vertex form, address the following:

a. What are the coordinates of the vertex?

b. Does the graph "open up" or "open down"?

c. What is the equation of the axis of symmetry?

d. Find any x-intercepts.

e. Find the y-intercept.

f. Sketch the graph.

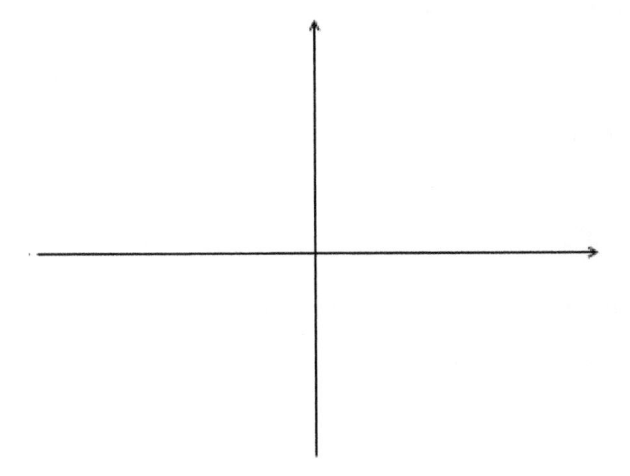

g. State the domain and range in interval notation.

Section 4.1 Objective 3 Graphing Quadratic Functions by Completing the Square

Work through the video that accompanies Example 3:

Rewrite the quadratic function $f(x) = 2x^2 - 4x - 3$ in vertex form, and then answer the questions below.

a. What are the coordinates of the vertex?

b. Does the graph "open up" or "open down"?

c. What is the equation of the axis of symmetry?

d. Find any x-intercepts.

e. Find the y-intercept.

f. Sketch the graph.

g. State the domain and range in interval notation.

Section 4.1 Objective 4 Graphing Quadratic Functions Using the Vertex Formula

Watch the video that accompanies Objective 4 and write your notes here:

Formula for the Vertex of a Parabola (Fill in the blanks)

Given a quadratic function of the form $f(x) = ax^2 + bx + c,\ a \neq 0,$ the vertex of the parabola is

______________________________ .

Work through Example 4:

Given the quadratic function $f(x) = -2x^2 - 4x + 5$, address the following:

a. What are the coordinates of the vertex?

b. Does the graph "open up" or "open down"?

c. What is the equation of the axis of symmetry?

d. Find any x-intercepts.

e. Find the y-intercept.

f. Sketch the graph.

g. State the domain and range in interval notation.

Section 4.1 Objective 5 Determining the Equation of a Quadratic Function Given Its Graph

Work through the video that accompanies Example 5: Analyze the graph to address the following about the quadratic function it represents.

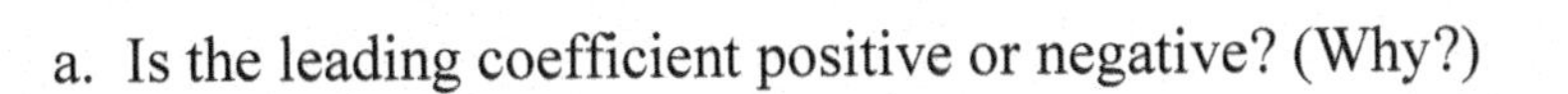

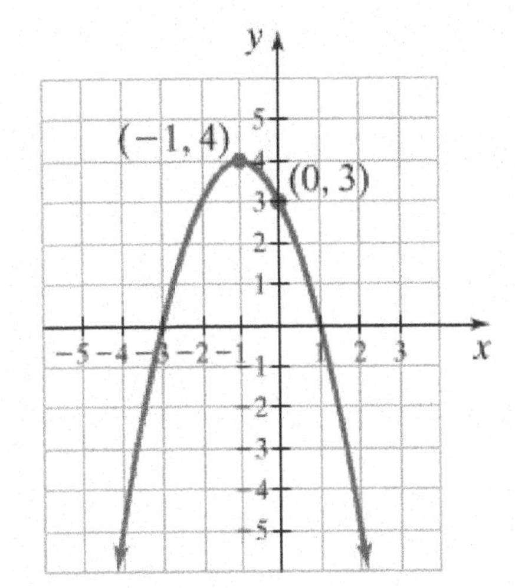

a. Is the leading coefficient positive or negative? (Why?)

b. What is the value of h? What is the value of k?

c. What is the value of the leading coefficient, a?

d. Write the equation of the function in vertex form
$f(x) = a(x-h)^2 + k$.

e. Write the equation of the function in the form
$f(x) = ax^2 + bx + c$.

Section 4.2 Guided Notebook

Section 4.2 Applications and Modeling of Quadratic Functions

- ☐ Work through Objective 1
- ☐ Work through Objective 2
- ☐ Work through Objective 3

Section 4.2 Applications and Modeling of Quadratic Functions

Work through the Introduction Video and take notes here: (Make sure that you write down the formula for the vertex of a quadratic function!)

Section 4.2 Objective 1 Maximizing Projectile Motion Functions

Work through the video that accompanies Example 1:

A toy rocket is launched with an initial velocity of 44.1 meters per second from a 1-meter-tall platform. The height h of the object at any time t seconds after launch is given by the function $h(t) = -4.9t^2 + 44.1t + 1$. How long after launch did it take the rocket to reach its maximum height? What is the maximum height obtained by the toy rocket?

Work through the video that accompanies Example 2: If an object is launched at an angle of 45 degrees from a 10-foot platform at 60 feet per second, it can be shown that the height of the object in feet is given by the quadratic function

$h(x) = -\frac{32x^2}{(60)^2} + x + 10$, where x is the horizontal distance of the object from the platform.

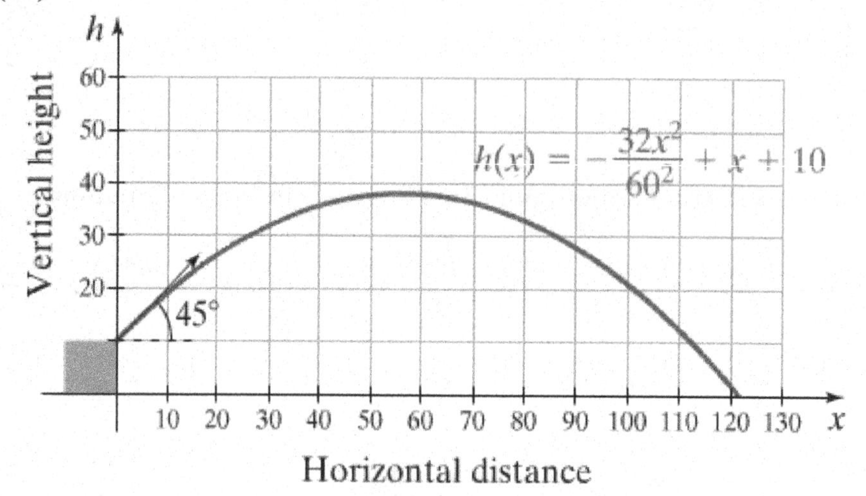

a. What is the height of the object when its horizontal distance from the platform is 20 feet? Round to two decimal places.

b. What is the horizontal distance from the platform when the object is at its maximum height?

c. What is the maximum height of the object?

Section 4.2 Objective 3 Maximizing Area Functions

Suppose that you have 3,000 feet of fencing to construct the rectangular pen that borders a river as seen below. Watch the first video located at the bottom of p. 4.2-19 to see a couple of different ways that you can construct this fence. What should the length and width of this fence be in order to maximize area? What is the maximum area?

Work through the video that accompanies Example 6:

Mark has 100 feet of fencing available to build a rectangular pen for his hens and roosters. One side of the pen will border a river and requires no fencing. He wants to separate the hens and roosters by dividing the pen into two equal areas. Let x represent the length of the center partition that is perpendicular to the river.

a. Create a function $A(x)$ that describes the total area of the rectangular enclosure as a function of x where x is the length of the center partition.

b. Find the length of the center partition that will yield the maximum area.

c. Find the length of the side of the fence parallel to the river that will yield the maximum area.

d. What is the maximum area?

Section 4.3 Guided Notebook

Section 4.3 The Graphs of Polynomial Functions

- ▯ Work through Objective 1
- ▯ Work through Objective 2
- ▯ Work through Objective 3
- ▯ Work through Objective 4
- ▯ Work through Objective 5
- ▯ Work through Objective 6
- ▯ Work through Objective 7

Section 4.3 The Graphs of Polynomial Functions

Section 4.3 Objective 1 Understanding the Definition of a Polynomial Function

Work through the video that accompanies Objective 1 and take notes here:

What is the definition of a **polynomial function**?

Determine which functions are polynomial functions. If the function is a polynomial function, identify the degree, the leading coefficient, and the constant coefficient.

$$f(x) = \sqrt{3}x^3 - 2x^2 - \frac{1}{6}$$

$$g(x) = 4x^5 - 3x^3 + x^2 + \frac{7}{x} - 3$$

Now work through Example 1 (Part c): Determine if the function is a polynomial function. If the function is a polynomial function, identify the degree, the leading coefficient, and the constant coefficient.

$$h(x)=\frac{3x-x^2+7x^4}{9}$$

Section 4.3 Objective 2 Sketching the Graphs of Power Functions

The graphs of five power functions are seen here.

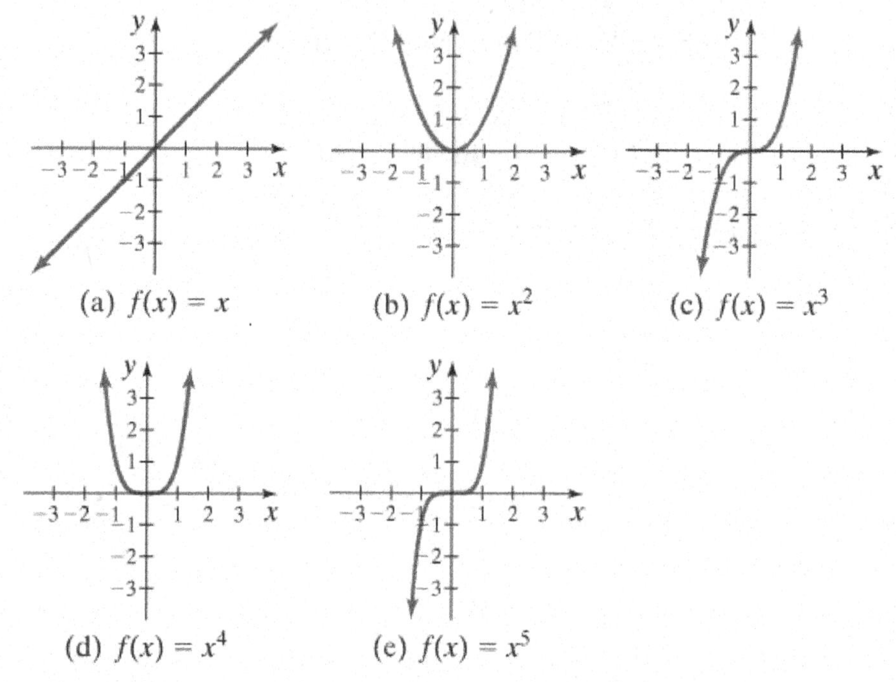

(a) $f(x)=x$ (b) $f(x)=x^2$ (c) $f(x)=x^3$

(d) $f(x)=x^4$ (e) $f(x)=x^5$

Work through the interactive video that accompanies Example 2 and use a power function to sketch the following functions:

a. $f(x)=-x^6$

b. $f(x) = (x+1)^5 + 2$

c. $f(x) = 2(x-3)^4$

Section 4.3 Objective 3 Determining the End Behavior of Polynomial Functions

Work through the video that accompanies Objective 3 and explain how we determine the end behavior of the graph of a polynomial function.

Write down the Two-Step Process for Determining the End Behavior of a Polynomial Function.

Step 1:

Step 2:

Work through the video that accompanies Example 3: Use the end behavior of each graph to determine whether the degree is even or odd and whether the leading coefficient is positive or negative.

a.

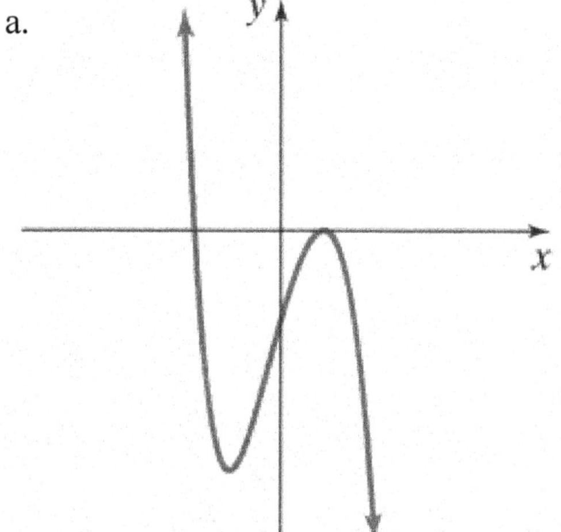

b.

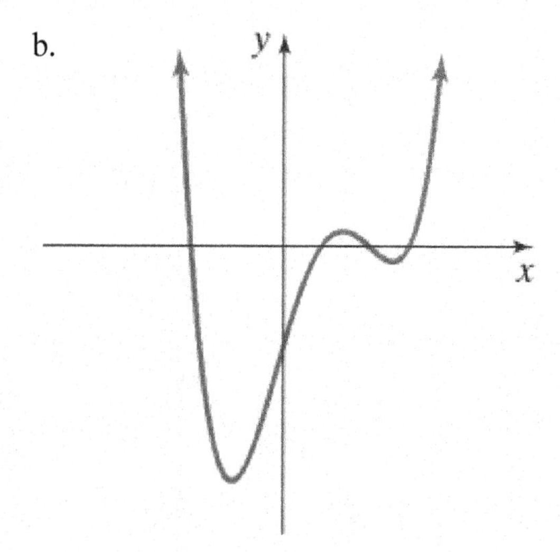

Section 4.3 Objective 4 Determining the Intercepts of a Polynomial Function

Read the introduction to this objective.

Explain how to determine the y-intercept of a polynomial function:

Explain how to find the x-intercepts (also called **real zeros**) of a polynomial function:

Work through Example 4:

Find the intercepts of the polynomial function $f(x) = x^3 - x^2 - 4x + 4$.

Now that you have found the intercepts of the function $f(x) = x^3 - x^2 - 4x + 4$, determine the end behavior of this function and start thinking about what this graph is starting to look like.

Section 4.3 Objective 5 Determining the Real Zeros of Polynomial Functions and Their Multiplicities

Work through the video that accompanies Objective 5 and take notes and fill in all blanks below:

If $(x-c)$ is a factor of a polynomial function, then ____________________.
(fill in the blank)

Multiplicity

If $(x-c)^k$ is a factor of a polynomial function, where k is a positive integer, then

____.
(fill in the blank)

If $(x-c)^k$ is a factor of a polynomial function, where c is a real number and k is an integer such that $k \geq 1$, then

- If k is odd, then ____

- If k is even, then ____

Work through the last part of the video that accompanies Example 5:
Find all real zeros of $f(x) = x(x^2 - 1)(x - 1)$. Determine the multiplicities of each zero, and decide whether the graph touches or crosses at each zero. (Try to create a rough sketch of this graph.)

Section 4.3 Objective 6 Sketching the Graph of a Polynomial Function

Carefully read through this objective and answer the following questions:

Can you rewrite the function $f(x) = x^3 - x^2 - 4x + 4$ in factored form? What is the factored form of this function?

What are the three zeros and their multiplicities of the function $f(x) = x^3 - x^2 - 4x + 4$? What is the y-intercept of this function?

Now plot the three zeros and plot the y-intercept. Also use the end behavior to begin to create a rough sketch of the function.

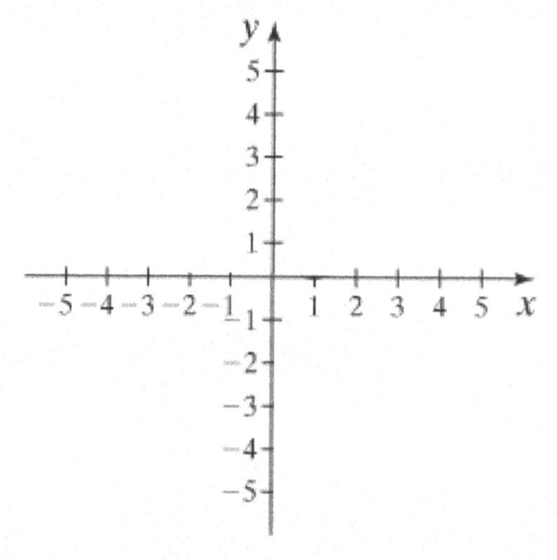

What is the definition of a **test value**?

Choose two test values as described in the eText then complete a “rough sketch” of $f(x) = x^3 - x^2 - 4x + 4$ on the grid above.

What is the definition of a **turning point**?

How many turning points can the graph of a polynomial of degree n have?

What is a turning point in which the graph changes from increasing to decreasing called?

What is a turning point in which the graph changes from decreasing to increasing called?

Write down the **Four-Step Process for Sketching the Graph of Polynomial Functions.**

Step 1.

Step 2.

Step 3.

Step 4.

Work through the interactive video that accompanies Example 6:

Use the four-step process to sketch the graphs of the following polynomial functions.

a. $f(x) = -2(x+2)^2(x-1)$

b. $f(x) = x^4 - 2x^3 - 3x^2$

Section 4.3 Objective 7 Determining a Possible Equation of a Polynomial Function Given Its Graph

Work through the Interactive Video that accompanies Example 7:
Analyze the graph to address the following about the polynomial function it represents.

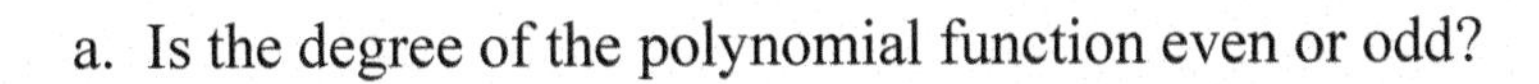

a. Is the degree of the polynomial function even or odd?

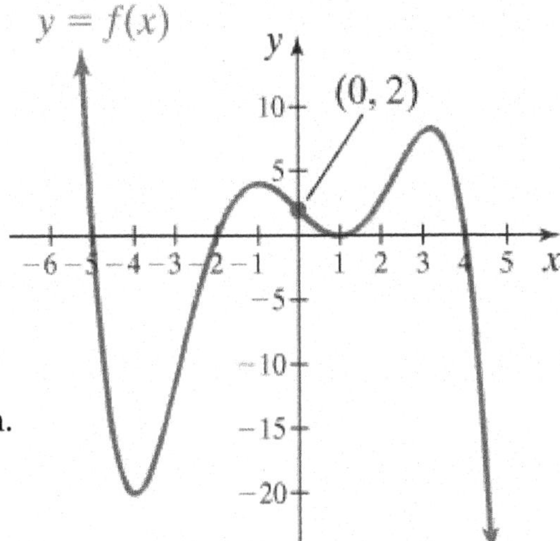

b. Is the leading coefficient positive or negative?

c. What is the value of the constant coefficient?

d. Identify the real zeros, and state the multiplicity of each.

e. Select from this list a possible function that could be represented by this graph.

i. $f(x) = -\frac{1}{20}(x+5)(x+2)(x-1)^2(x-4)$

ii. $f(x) = -\frac{1}{800}(x+5)^2(x+2)^2(x-1)(x-4)^2$

iii. $f(x) = \frac{1}{20}(x+5)(x+2)(x-1)^2(x-4)$

iv. $f(x) = -\frac{1}{10}(x+5)(x+2)(x-1)^2(x-4)$

Section 4.4 Guided Notebook

Section 4.4 Synthetic Division; The Remainder and Factor Theorems

- ☐ Work through Section 4.1 TTK #1
- ☐ Work through Section 4.1 TTK #3
- ☐ Work through Section 4.1 TTK #5
- ☐ Work through Objective 1
- ☐ Work through Objective 2
- ☐ Work through Objective 3
- ☐ Work through Objective 4
- ☐ Work through Objective 5

Section 4.4 Synthetic Division; The Remainder and Factor Theorems

Section 4.4 Objective 1 Using the Division Algorithm

Write down the **Division Algorithm.**

Work through Example 1: Given the polynomials $f(x) = 4x^4 - 3x^3 + 5x - 6$ and $d(x) = x - 2$, find polynomials $q(x)$ and $r(x)$, and write $f(x)$ in the form $f(x) = d(x)q(x) + r(x)$.

Write down the **Corollary to the Division Algorithm.**

Section 4.4 Objective 2 Using Synthetic Division

Synthetic division is a very handy way to long divide polynomials when the divisor is of the form $x-c$. Carefully work through the video that accompanies Objective 2 and **use synthetic division** to divide $f(x)=4x^4-3x^3+5x-6$ by $x-2$.

Work through the video that accompanies Example 2: Use synthetic division to divide $f(x)$ by $x-c$, and then write $f(x)$ in the form $f(x)=(x-c)q(x)+r$ for $f(x)=-2x^4+3x^3+7x^2-x+5$ divided by $x+1$.

Section 4.4 Objective 3 Using the Remainder Theorem

Carefully read through this objective and take notes here:

Write down **The Remainder Theorem.**

Work through the video that accompanies Example 3:

Use the remainder theorem to find the remainder when $f(x)$ is divided by $x-c$.

a. $f(x)=5x^4-8x^2+3x-1;\ x-2$

b. $f(x)=3x^3+5x^2-5x-6;\ x+2$

Section 4.4 Objective 4 Using the Factor Theorem

What is the definition of a **factor**?

What does it mean for a polynomial to be a factor of another polynomial?

Write down the **Factor Theorem.**

Work through the video that accompanies Example 4:

Determine whether $x+3$ is a factor of $f(x)=2x^3+7x^2+2x-3$.

Section 4.4 Objective 5 Sketching the Graph of a Polynomial Function

Carefully read through this objective and show how to completely factor the polynomial $f(x) = 2x^3 + 7x^2 + 2x - 3$.

$f(x) =$ ______________________________

(Write $f(x) = 2x^3 + 7x^2 + 2x - 3$ in completely factored form.)

Once you have shown how to completely factor $f(x) = 2x^3 + 7x^2 + 2x - 3$, click on the video link to see how to use the four-step process to sketch the graph of $f(x) = 2x^3 + 7x^2 + 2x - 3$.

Work through the video that accompanies Example 5:
Given that $x = 2$ is a zero of $f(x) = x^3 - 6x + 4$, completely factor f and sketch its graph.
(Be sure to write f in completely factored form.)

Section 4.5 Guided Notebook

Section 4.5 The Zeros of Polynomial Functions; The Fundamental Theorem of Algebra

- ☐ Work through Objective 1
- ☐ Work through Objective 2
- ☐ Work through Objective 3
- ☐ Work through Objective 4
- ☐ Work through Objective 5
- ☐ Work through Objective 6
- ☐ Work through Objective 7

Section 4.5 The Zeros of Polynomial Functions; The Fundamental Theorem of Algebra

Section 4.5 Objective 1 Using the Rational Zeros Theorem

Look at your previous pages of notes and see that we were given that $x=2$ was a zero of $f(x)=x^3-6x+4$. We used this information to completely factor *f* as $f(x)=(x-2)\left(x-(-1+\sqrt{3})\right)\left(x-(-1-\sqrt{3})\right)$. What if we were not given the fact that $x=2$ was a zero? How would we have found the three zeros? The answer is that we first must use the **rational zeros theorem.**

What is the definition of a **rational zero**?

Write down the **Rational Zeros Theorem.**

Work through Example 1: Use the rational zeros theorem to determine the potential zeros of the polynomial function $f(x) = 4x^4 - 7x^3 + 9x^2 - x - 10$.

<u>Section 4.5 Objective 2 Using Descartes' Rule of Signs</u>

Write down Descartes' Rule of Signs in the box below:

Let f be a polynomial function with real coefficients written in descending order.

1.

2.

Work through Example 2: Determine the possible number of positive real zeros and negative real zeros of each polynomial function.

a. $f(x) = 2x^5 + 3x^4 + 8x^2 + 2x$

b. $f(x) = 6x^4 + 13x^3 + 61x^2 + 8x - 10$

Section 4.5 Objective 3 Finding the Zeros of a Polynomial Function

Write down the **Fundamental Theorem of Algebra.**

Write down the **Number of Zeros Theorem.**

Work through the video that accompanies Example 3: Find all complex zeros of $f(x)=6x^4+13x^3+61x^2+8x-10$ and rewrite $f(x)$ in completely factored form.

TIP: Once we factor a polynomial into the product of linear factors and a quadratic function of the form $f(x)=(x-c_1)(x-c_2)(x-c_3)\cdots(x-c_k)\left(ax^2+bx+c\right)$, we need simply to solve the quadratic equation $ax^2+bx+c=0$ to find the remaining two zeros. For example, if we want to find all zeros of $g(x)=2x^3-3x^2+4x-3$, we need only try to find **one** zero from the list created by the Rational Zeros Theorem. Then find the two zeros of the remaining quadratic function.

See if you can find the zeros of $g(x)=2x^3-3x^2+4x-3$. Watch the **video** by clicking on the word "video" seen in your eText at the end of the last paragraph of this objective.

Section 4.5 Objective 4 Solving Polynomial Equations

Work through the video that accompanies Example 4:

Solve $5x^5 - 9x^4 + 23x^3 - 35x^2 + 12x + 4 = 0.$

Section 4.5 Objective 5 Using the Complex Conjugate Pairs Theorem

Write down the **Complex Conjugate Pairs Theorem.**

Work through the interactive video that accompanies Example 5:
Find the equation of a polynomial function *f* with real coefficients that satisfies the given conditions:

a. Fourth-degree polynomial function such that 1 is a zero of multiplicity 2 and $2-i$ is also a zero.

Example 5 b. Find the fifth-degree polynomial function sketched below given that $1+3i$ is a zero.

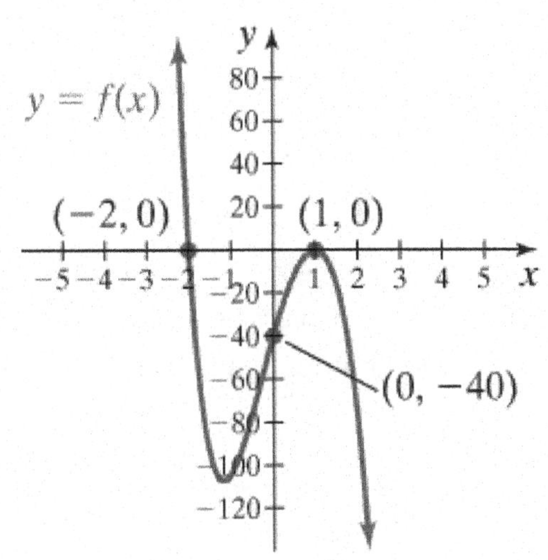

Section 4.5 Objective 6 Using the Intermediate Value Theorem

Why is it true that every odd degree polynomial with real coefficients has at least one real zero?

Write down the **Intermediate Value Theorem.** Also create a graph that depicts the Intermediate Value Theorem. (You can use the Figure that is seen in the eText just below the theorem.)

Work through the video that accompanies Example 6:

Use the Intermediate Value Theorem to find the real zero of $f(x) = x^3 + 2x - 1$ correct to two decimal places.

Section 4.5 Objective 7 Sketching the Graphs of Polynomial Functions

Write down the **Steps for Sketching the Graphs of Polynomial Functions.**

Step 1.

Step 2.

Step 3.

Step 4.

Work through the video that accompanies Example 7:

Sketch the graph of $f(x)=2x^5-5x^4-2x^3+7x^2-4x+12$.

Section 4.6 Guided Notebook

Section 4.6 Rational Functions and Their Graphs

- ☐ Work through TTK# 1
- ☐ Work through TTK# 2
- ☐ Work through TTK# 3
- ☐ Work through Objective 1
- ☐ Work through Objective 2
- ☐ Work through Objective 3
- ☐ Work through Objective 4
- ☐ Work through Objective 5
- ☐ Work through Objective 6
- ☐ Work through Objective 7

Section 4.6 Rational Functions and Their Graphs

Introduction to Section 4.6

Write down the definition of a **Rational Function**.

Section 4.6 Objective 1: Finding the Domain and Intercepts of Rational Functions

What is the domain of a rational function?

How do you find the y-intercept of a rational function if there is one?

How do you find the x-intercepts of a rational function?

Work Example 1 showing all work below. To see the steps of the solution watch the video with the example.

Let $f(x)=\dfrac{x-4}{x^2+x-6}$.

a. Determine the domain of f.

b. Determine the y-intercept (if any).

c. Determine any x-intercepts.

Section 4.6 Objective 2: Identifying Vertical Asymptotes

Sketch the reciprocal function, $f(x)=\frac{1}{x}$.

Look at the graph of $f(x)=\frac{1}{x}$ and fill in the blanks below:

The values of f approach _______ as x approaches ________ from the right.

The values of f approach _______ as x approaches ________ from the left.

Write down the definition of a **vertical asymptote** and sketch the four graphs as seen in the definition.

What is the caution statement say about locating vertical asymptotes?

Watch the video with Example 2 and answer the following questions.

Find the vertical asymptotes (if any) of the function $f(x) = \dfrac{x-3}{x^2+x-6}$ and then sketch the graph near the vertical asymptotes.

1. What is the first step?

2. What are the vertical asymptotes?

3. When graphing vertical asymptotes the lines must be ____________________?

4. How do you sketch the graph near the asymptotes?

5. Show the work for each of the test values below. Include a sketch.

6. What is the side note about choosing the values and the x-intercept?

Work through the video with Example 3 and take notes below.
Find the vertical asymptotes (if any) of the following function and then sketch the graph near the vertical asymptotes.

$$f(x)=\frac{x+3}{x^2+x-6}$$

1. What is the first step?

2. What happens to the common factor?

3. What is the vertical asymptote?

4. Show all steps to sketch the graph near the asymptote.

Section 4.6 Objective 3: Identifying Horizontal Asymptotes

Watch the video with Objective 3 and answer the following questions.

1. Sketch the graph of the reciprocal function $f(x)=\frac{1}{x}$.

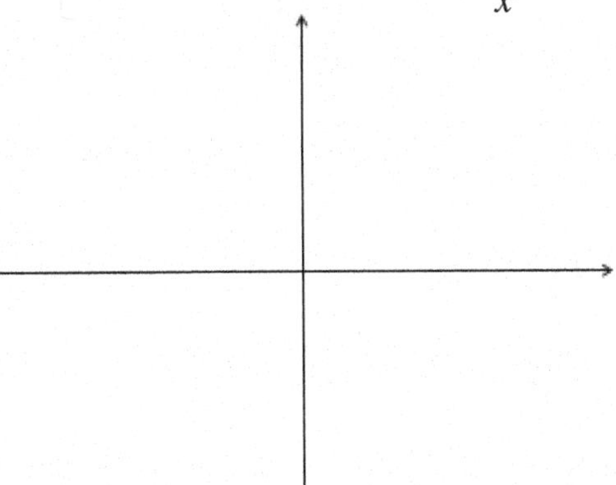

2. What do you notice about the graph and the y-axis?

3. What do you notice about the graph and the x-axis?

4. What is the x-axis?

5. Write the definition of a **Horizontal Asymptote**.

6. Draw the illustration for the first 2 properties of **Horizontal Asymptotes of Rational Functions**.

Write down the three Properties of **Horizontal Asymptotes of Rational Functions**.
(Page 4.6-19)

Finding Horizontal Asymptotes of a Rational Function

Let $f(x) = \frac{g(x)}{h(x)} = \frac{a_n x^n + a_{n-1}x^{n-1} + a_{n-2}x^{n-2} + \cdots + a_1 x + a_0}{b_m x^m + b_{m-1}x^{m-1} + b_{m-2}x^{m-2} + \cdots + b_1 x + b_0}$, $a_n \neq 0$, $b_m \neq 0$, where f is written in lowest terms, n is the degree of g, and m is the degree of h.

- If $m > n$, then $y = 0$ is the horizontal asymptote.
- If $m = n$, then the horizontal asymptote is $y = \frac{a_n}{b_m}$, the ratio of the leading coefficients.
- If $m < n$, then there are no horizontal asymptotes.

Work through Example 4 and show all work below. For detailed step by step solutions watch the video.

Find the horizontal asymptote of the graph of each rational function or state that one does not exist.

a. $f(x) = \dfrac{x}{x^2 - 4}$

b. $f(x) = \dfrac{4x^2 - x + 1}{1 - 2x^2}$

c. $f(x) = \dfrac{2x^3 + 3x^2 - 2x - 2}{x - 1}$

Section 4.6 Objective 4: Using Transformations to Sketch the Graphs of Rational Functions.

Write down the properties of $f(x)=\frac{1}{x}$ and $f(x)=\frac{1}{x^2}$.

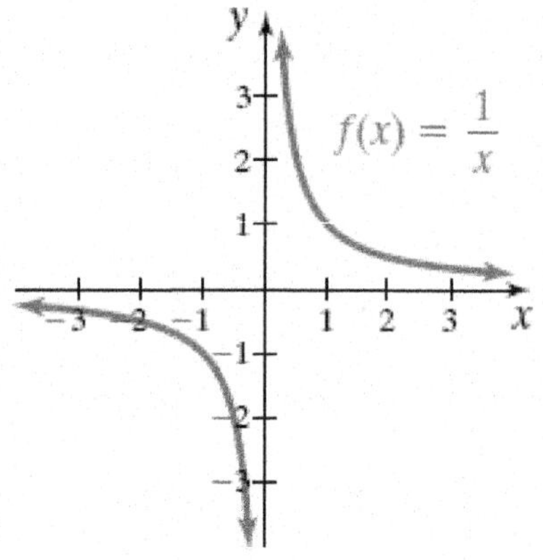

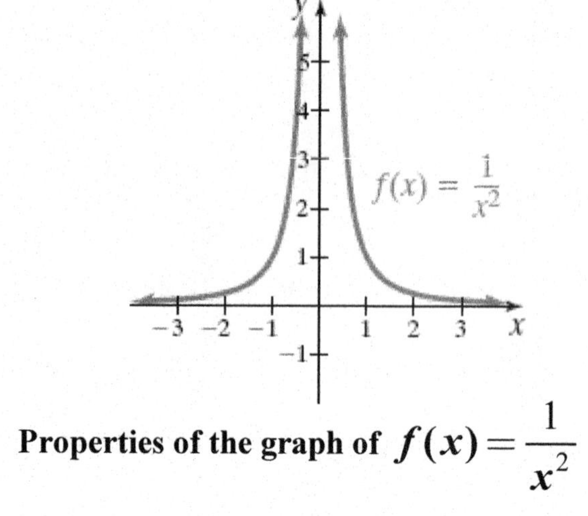

Properties of the graph of $f(x)=\frac{1}{x}$

Properties of the graph of $f(x)=\frac{1}{x^2}$

Work through Example 5 showing all work below. The video will show the step by step solution.

Use transformations to sketch the graph of $f(x) = \frac{-2}{(x+3)^2} + 1$.

Section 4.6 Objective 5: Sketching Rational Functions Having Removable Discontinuities

What are removable discontinuities? When do they occur?

Work through Example 6 showing all work below. For a detailed step by step solution watch the video.

Sketch the graph of the following function and find the coordinates of all removable discontinuities. $f(x)=\dfrac{x^2-1}{x+1}$.

Work through Example 7 showing all work below. For a detailed step-by-step solution, watch the video with the example.

Sketch the graph of the following function and find the coordinates of all removable discontinuities. $f(x)=\dfrac{x+3}{x^2+x-6}$.

What is the caution statement about graphing calculators and removable discontinuities?

Section 4.6 Objective 6: Identifying Slant Asymptotes

When does a slant asymptote occur?

How do you find the slant asymptote?

Watch the video with Example 8 and take notes here:

Find the slant asymptote of $f(x)=\dfrac{2x^2+3x-2}{x-1}$.

Section 4.6 Objective 7: Sketching Rational Functions

What are the steps for Graphing Rational Functions of the Form $f(x)=\frac{g(x)}{h(x)}$?

Step 1.

Step 2.

Step 3.

Step 4.

Step 5.

Step 6.

Step 7.

Step 8.

Step 9.

Watch the video with Example 9 answering the following questions.

Sketch the graph of $f(x)=\dfrac{x^3+2x^2-9x-18}{x^3+6x^2+5x-12}$.

1. Write down the steps to find the domain.

2. After finding the domain, what is the next step? Show all work for this step below.

3. What is the simplified function?

4. What happens at $x=-3$?

5. Is the function even or odd? What does that indicate about the graph?

6. Why is it a good idea to test for symmetry?

7. Show the steps to find the intercepts below.

8. Which denominator is used to find the vertical asymptotes? What are they?

9. Verify the behavior near the vertical asymptotes below.

10. Will there be a horizontal or slant asymptote? Why?

11. What is the horizontal or slant asymptote? How was it found?

12. How do you determine which values of x to evaluate?

13. Show the steps to evaluate the function at these values.

14. Draw the final sketch of the graph below.

Section 5.1 Guided Notebook

Section 5.1 Exponential Functions

- ☐ Work through Section 5.1 TTK #1
- ☐ Work through Section 5.1 TTK #2
- ☐ Work through Objective 1
- ☐ Work through Objective 2
- ☐ Work through Objective 3
- ☐ Work through Objective 4

Section 5.1 Exponential Functions

Section 5.1 Objective 1 Understanding the Characteristics of Exponential Functions

Write down the **definition of an exponential function.**

Click on the video link found in this objective and take notes here: **Write down the charactericstics of the exponential function as seen near the end of this video.**

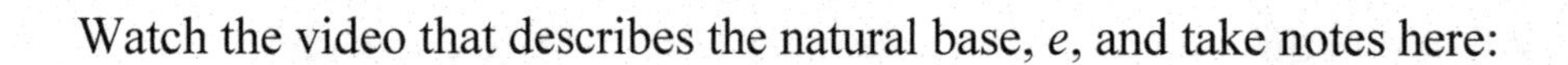

Watch the video that describes the natural base, *e*, and take notes here:

What is the number *e* rounded to 6 decimal places?

Sketch the graphs of $y = 2^x$, $y = 3^x$, and $y = e^x$ on the same grid.

Make sure that you have a scientific calculator that has an $\boxed{e^x}$ key. Then do Example 1: Evaluate each expression correctly to six decimal places.

a. $\left(\frac{3}{7}\right)^{0.6}$

b. $9\left(\frac{5}{8}\right)^{-0.375}$

c. e^2

d. $e^{-0.534}$

e. $1000e^{0.013}$

Work through the interactive video that accompanies Example 2 and take notes here:

Sketch the graph of each exponential function.

a. $f(x)=\left(\frac{2}{3}\right)^x$

b. $f(x)=-2e^x$.

Work through the video that accompanies Example 3 and take notes here:

Find the exponential function $f(x) = b^x$ whose graph is given as follows.

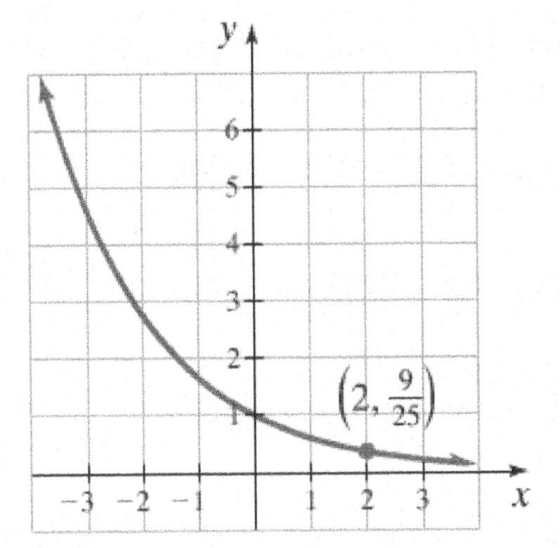

Section 5.1 Objective 2 Sketching the Graphs of Exponential Functions Using Transformations

Sketch the graphs of $y = 3^x$ and $y = 3^x - 1$ here:

Work through the video that accompanies Example 4 and take notes here:

Use transformations to sketch the graph of $f(x) = -2^{x+1} + 3$.

(Make sure to label the three points as shown in the video on your final graph.)

Work through the video that accompanies Example 5 and take notes here:

Use transformations to sketch the graph of $f(x) = e^{-x} - 2$. Determine the domain, range, y-intercept, and find the equation of any asymptotes. (Make sure to label the three points as shown in the video on your final graph.)

Take some time to experiment with the **Guided Visualization** titled "Sketching Exponential Functions" seen on page 5.1-21. Sketch two exponential functions below. One function should be base e and the other function should be a base other than e. Clearly label your functions and give the equations of the vertical asymptotes.

Section 5.1 Objective 3 Solving Exponential Equations by Relating the Bases

Why is every exponential function of the form $f(x) = b^x$ one-to-one?

A function f is one-to-one if for any two range values $f(u)$ and $f(v)$,

$f(u) = f(v)$ implies that _______________.

(Fill in the Blank)

Method of Relating the Bases

If an exponential function can be written in the form $b^u = b^v$, then

____________________.

(Fill in the Blank)

Carefully work through the **animation** that accompanies Example 6 and take notes here:

Solve the following equations:

a. $8 = \frac{1}{16^x}$

b. $\frac{1}{27^x} = \left(\sqrt[4]{3}\right)^{x-2}$

Work through the interactive video that accompanies Example 7.

Use the method of relating the bases to solve each exponential equation.

a. $e^{3x-1} = \frac{1}{\sqrt{e}}$

b. $\frac{e^{x^2}}{e^{10}} = \left(e^{x}\right)^3$

Section 5.1 Objective 4 Solving Applications of Exponential Functions

Work through Example 8:

Most golfers find that their golf skills improve dramatically, at first, and then level off rather quickly. For example, suppose that the distance (in yards) that a typical beginning golfer can hit a 3-wood after t weeks of practice on the driving range is given by the exponential function $d(t) = 225 - 100e^{-0.7t}$. This function has been developed after many years of gathering data on beginning golfers.

How far can a typical beginning golfer initially hit a 3-wood? Round to the nearest hundredth yard.

How far can a typical beginning golfer hit a 3-wood after 1 week of practice on the driving range? Round to the nearest hundredth yard.

After 5 weeks? Round to the nearest hundredth yard.

After 9 weeks? Round to the nearest hundredth yard.

Compound Interest

What is simple interest?

Read through the portion of your eText that describes compound interest to see how the formula for periodic compound interest is derived.

Write the **Periodic Compound Interest Formula** here:

For interest compounded…

annually, use $n = 1$
semi-annually, use $n = 2$
quarterly, use $n = 4$
monthly, use $n = 12$
daily, use $n = 365$

Work through the video that accompanies Example 9:

Which investment will yield the most money after 25 years?

Investment A: $12,000 invested at 3% compounded monthly.
Investment B: $10,000 invested at 3.9% compounded quarterly.

Continuous Compound Interest

Watch the animation seen in the portion of your eText that describes continuous compound interest to see how to derive the formula for continuous compound interest.

Write the **Continuous Compound Interest Formula** here:

Work through the video that accompanies Example 10:
How much money would be in an account after 5 years if an original investment of $6,000 was compounded continuously at 4.5%? Compare this amount to the same investment that was compounded daily.

Present Value for Periodic Compound Interest

Present value is the amount of money needed now (in the present) to reach an investment goal in the future.

Show how to derive the formula for present value for periodic compound interest by starting with the formula for periodic compound interest here:

Write down the **Present Value Formula for Periodic Compound Interest** here:

Write down the **Present Value Formula for Continuous Compound Interest** here:

Work through the video that accompanies Example 11:

a. Find the present value of $8000 if interest is paid at a rate of 5.6% compounded quarterly for 7 years. Round to the nearest cent.

b. Find the present value of $18,000 if interest is paid at a rate of 8% compounded continuously for 20 years. Round to the nearest cent.

Exponential Growth Model

Write down the exponential growth model.

Work through the video that accompanies Example 12:

The population of a small town follows the exponential growth model $P(t) = 900e^{.015t}$, where t is the number of years after 1900. Answer the following questions, rounding each answer to the nearest whole number.

a. What was the population of this town in 1900?

b. What was the population of this town in 1950?

c. Use this model to predict the population of this town in 2012.

Work through the video that accompanies Example 13:

Twenty years ago, the State of Idaho Fish and Game Department introduced a new breed of wolf into a certain Idaho forest. The current wolf population in this forest is now estimated at 825, with a relative growth rate of 12%. Answer the following questions, rounding each answer to the nearest whole number.

a. How many wolves did the Idaho Fish and Game Department introduce into this forest?

b. How many wolves can be expected after another 20 years?

Section 5.2 Guided Notebook

Section 5.2 Logarithmic Functions

- ☐ Work through Section 5.2 TTK #3
- ☐ Work through Section 5.2 TTK #4
- ☐ Work through Section 5.2 TTK #6
- ☐ Work through Section 5.2 TTK #9
- ☐ Work through Objective 1
- ☐ Work through Objective 2
- ☐ Work through Objective 3
- ☐ Work through Objective 4
- ☐ Work through Objective 5
- ☐ Work through Objective 6
- ☐ Work through Objective 7

Section 5.2 Logarithmic Functions

Section 5.2 Objective 1 Understanding the Definition of a Logarithmic Function

Work through the video that accompanies Objective 1 and take notes here:

Sketch the graph of $f(x)=b^x$, $b>1$ as seen in the video and plot several points that lie on the graph, then sketch the graph of the inverse function.

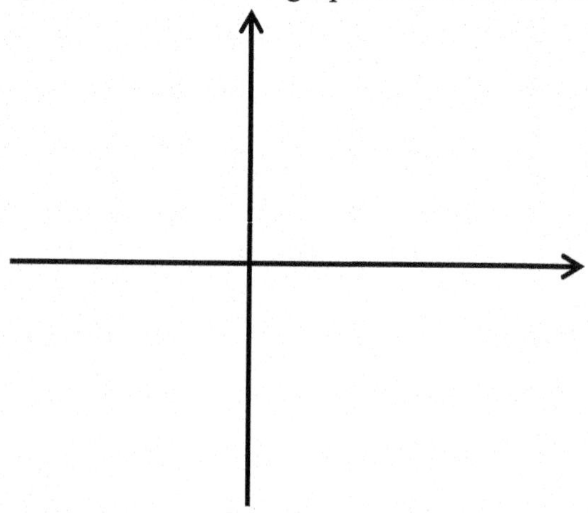

We want to find the equation of the inverse of $f(x)=b^x$. Follow the 4-step process for finding inverse functions, as stated in this video, and see if we can find the equation of this inverse function.

Step 1:

Step 2:

Step 3:

Before we can complete Step 3, we must define the logarithmic function. Write the definition of a logarithmic function here:

Definition of the Logarithmic Function

Now finish Step 3:

Step 4:

Work through the video that accompanies Example 1:

Write each exponential equation as an equation involving a logarithm.

a. $2^3 = 8$

b. $5^{-2} = \frac{1}{25}$

c. $1.1^M = z$

Work through the video that accompanies Example 2:

Write each exponential equation as an equation involving an exponent.

a. $\log_3 81 = 4$

b. $\log_4 16 = y$

c. $\log_{3/5} x = 2$

Section 5.2 Objective 2 Evaluating Logarithmic Expressions

Read through Objective 2 in your eText and take notes here:

Explain how to evaluate the expression $\log_4 64$.

Write down "The Method of Relating the Bases."

Work through the interactive video that accompanies Example 3 and take notes here:
Evaluate each logarithm:

a. $\log_5 25$

b. $\log_3 \frac{1}{27}$

c. $\log_{\sqrt{2}} \frac{1}{4}$

Section 5.2 Objective 3 Understanding the Properties of Logarithms

Write down the two General Properties of Logarithms.

General Properties of Logarithms

For $b > 0$ and $b \neq 1$,

1.

2.

In Section 3.6, we studied one-to-one functions. Given a function f and the inverse function f^{-1}, we learned two composition cancellation equations. Write down the two composition cancellation equations here:

Composition Cancellation Equations

1.

2.

If $f(x) = b^x$ is an exponential function, the inverse function is $f^{-1}(x) = \log_b x$. Using this information and applying the two composition cancellation equations seen above, write down the two cancellation properties of exponentials and logarithms.

Cancellation Properties of Exponentials and Logarithms

For $b > 0$ and $b \neq 1$,

1.

2.

Work through Example 4:

Use the properties of logarithms that you have written above to evaluate each expression.

a. $\log_3 3^4$

b. $\log_{12} 12$

c. $7^{\log_7 13}$

d. $\log_8 1$

Section 5.2 Objective 4 Using the Common and Natural Logarithms

Read Objective 4 in your eText and take notes here. Then write down the definition of the **common logarithmic function** and the **natural logarithmic function.**

Common Logarithmic Function

Natural Logarithmic Function

Work through the video that accompanies Example 5:

Write each exponential equation as an equation involving a common logarithm or natural logarithm.

a. $e^0 = 1$

b. $10^{-2} = \frac{1}{100}$

c. $e^K = w$

Work through the video that accompanies Example 6:

Write each logarithmic equation as an equation involving an exponent.

a. $\log 10 = 1$

b. $\ln 20 = Z$

c. $\log(x-1) = T$

Work through the video that accompanies Example 7:

Evaluate each expression without the use of a calculator.

a. $\log 100$

b. $\ln \sqrt{e}$

c. $e^{\ln 51}$

d. $\log 1$

Section 5.2 Objective 5 Understanding the Characteristics of Logarithmic Functions

Write down the three steps to sketching the graph of a logarithmic function of the form $f(x) = \log_b x,\ b > 0,\ b \neq 1$.

Step 1:

Step 2:

Step 3:

Follow the 3 steps on the previous page to sketch the graph of $f(x) = \log_3 x$.

Work through the video that accompanies Example 8 to see how to sketch this function.

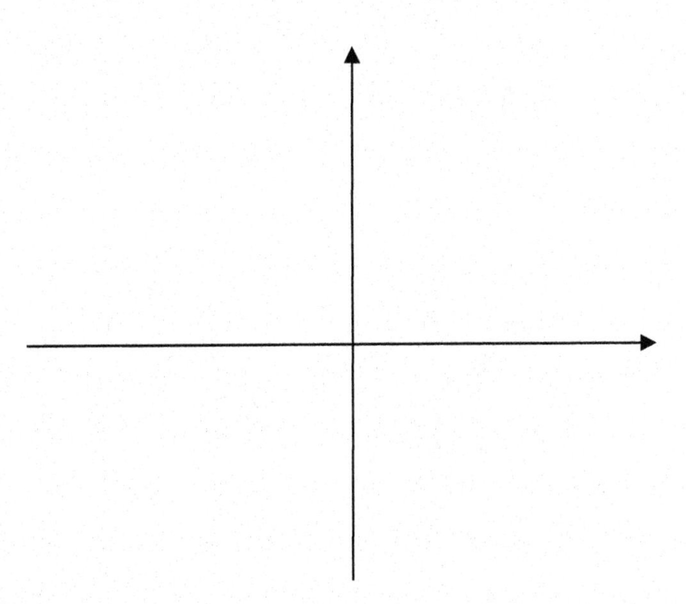

Write down the **Characteristics of Logarithmic Functions** as seen in your eText. Make sure to sketch the two graphs as seen on page 5.2-18 and plot the points, as shown on the graphs.

Section 5.2 Objective 6 Sketching the Graphs of Logarithmic Functions Using Transformations

Do you remember the transformation techniques (how to shift the graphs of functions) that were discussed in Section 3.4? You may want to review these techniques before working on the next set of exercises.

Work through the video that accompanies Example 9:
Sketch the graph of $f(x)=-\ln(x+2)-1$. (Make sure that your final graph contains the three points that are seen on the final graph as shown in the video.)

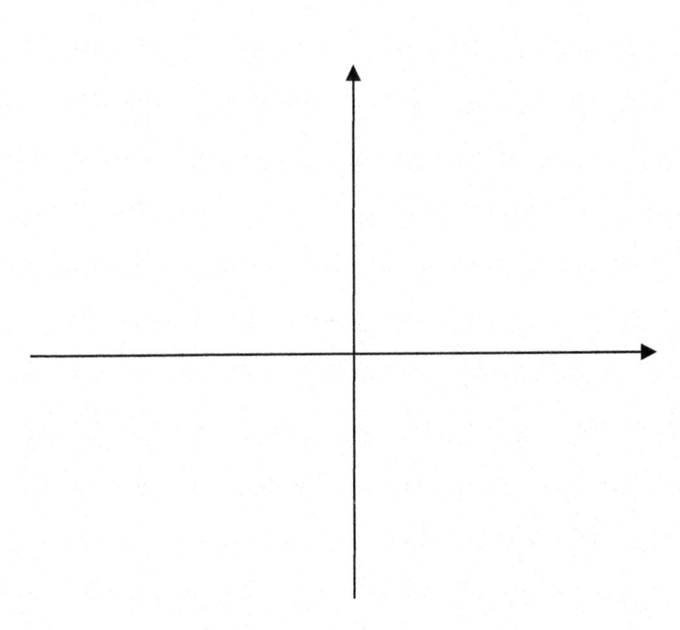

Take some time to experiment with the **Guided Visualization** titled "Sketching Logarithmic Functions" seen on page 5.2-21. Sketch two logarithmic functions below. One function should a natural logarithm and the other function should be a base other than e. Clearly label your functions and give the equations of the horizontal asymptotes.

Section 5.2 Objective 7 Finding the Domain of Logarithmic Functions

Look at your graph of $f(x) = -\ln(x+2) - 1$ from Example 9. What is the domain of this function? Do you see that the domain is $(-2, \infty)$? To find the domain of a logarithmic function of the form $f(x) = \log_b[g(x)]$, we must find the solution to the inequality $g(x) > 0$.

Example: Find the domain of $f(x) = \log_7[2x-5]$. To find the domain, we must solve the inequality $2x - 5 > 0$. Solving this inequality, we get $x > \frac{5}{2}$. So the domain of $f(x) = \log_7[2x-5]$ is $\left(\frac{5}{2}, \infty\right)$.

Now work through the interactive video that accompanies Example 10:

Find the domain of $f(x) = \log_5\left(\frac{2x-1}{x+3}\right)$.

Section 5.3 Guided Notebook

Section 5.3 Properties of Logarithms

- ☐ Work through Objective 1
- ☐ Work through Objective 2
- ☐ Work through Objective 3
- ☐ Work through Objective 4

Section 5.3 Properties of Logarithms

Section 5.3 Objective 1 Using the Product Rule, Quotient Rule, and Power Rule for Logarithms

In Section 5.2, you learned the following properties of logarithms:

General Properties of Logarithms
For $b > 0$ and $b \neq 1$,

1. $\log_b b = 1$
2. $\log_b 1 = 0$

Cancellation Properties of Exponentials and Logarithms
For $b > 0$ and $b \neq 1$,

1. $b^{\log_b b} = b$
2. $\log_b b^x = x$

It is now time to learn some more properties of logarithms. Write down the 3 properties of logarithms seen in Objective 1 of your eText. (You should watch the video proof of each property to get an understanding as to why these properties are true.)

Properties of Logarithms
If $b > 0$ and $b \neq 1$, u and v represent positive real numbers and r is any real number, then

1.

2.

3.

Work through the video that accompanies Example 1:

Use the product rule for logarithms to expand each expression. Assume $x > 0$.

a. $\ln(5x)$

b. $\log_2(8x)$

True or False: $\log_b(u+b) = \log_b u + \log_b v$

Work through the video that accompanies Example 2:

Use the quotient rule for logarithms to expand each expression. Assume $x > 0$.

a. $\log_5\left(\frac{12}{x}\right)$

b. $\ln\left(\frac{x}{e^5}\right)$

True or False: $\log_b(u-b) = \log_b u - \log_b v$

True or False: $\log_b u - \log_b v = \frac{\log_b u}{\log_b v}$

Work through the video that accompanies Example 3:

Use the power rule for logarithms to rewrite each expression. Assume $x > 0$.

a. $\log 6^3$

b. $\log_{1/2} \sqrt[4]{x}$

True or False: $(\log_b u)^r = r \log_b u$

Work through the video that accompanies Example 4:

Use properties of logarithms to evaluate each expression without the use of a calculator.

a. $7^{\log_7 6 + \log_7 3}$

b. $e^{2\ln 5 - \frac{1}{3}\ln 64 + \ln 1}$

Section 5.3 Objective 2 Expanding and Condensing Logarithmic Expressions

Work through the interactive video that accompanies Example 5:

Use properties of logarithms to expand each logarithmic expression as much as possible.

a. $\log_7\left(49x^3\sqrt[5]{y^2}\right)$

b. $\ln\left(\dfrac{\left(x^2-4\right)}{9e^{x^3}}\right)$

Work through the interactive video that accompanies Example 6:

Use properties of logarithms to rewrite each expression as a single logarithm.

a. $\dfrac{1}{2}\log(x-1)-3\log z+\log 5$

b. $\dfrac{1}{3}\left(\log_3 x-2\log_3 y\right)+\log_3 10$

Section 5.3 Objective 3 Solving Logarithmic Equations Using the Logarithm Property of Equality

Why are logarithmic functions one-to-one?

Write down the **Logarithm Property of Equality.**

Work through the interactive video that accompanies Example 7:

Solve the following equations:

a. $\log_7(x-1)=\log_7 12$

b. $2\ln x=\ln 16$

Explain why $x=-4$ is **not** a solution to $2\ln x=\ln 16$.

Section 5.3 Objective 4 Using the Change of Base Formula

Look at your scientific calculator and locate the [log] key and the [ln] key. If you are given an expression such as $\log 50$ or $\ln 319$, you can use your calculator to evaluate these expressions. See if you can use your calculator to evaluate these two logarithmic expressions. You should get $\log 50 \approx 1.69897$ and $\ln 319 \approx 5.76519$. How would you use your calculator to evaluate the expression $\log_3 10$? The answer is that you need to change the base from base 3 to base 10 or base e.

Write down the **Change of Base Formula.**

You have the skills necessary to prove the Change of Base Formula! Watch the video proof of the Change of Base Formula and write your notes here.

At the end of the video proof, see how to evaluate the expression $\log_3 10$.

$\log_3 10 =$

Work through Example 8:
Approximate the following expressions. Round each to four decimal places.

a. $\log_9 200$

b. $\log_{\sqrt{3}} \pi$

Work through the video that accompanies Example 9:

Use the change of base formula and the properties of logarithms to rewrite as a single logarithm involving base 2.

$$\log_4 x + 3\log_2 y$$

Work through the video that accompanies Example 10:

Use the change of base formula and the properties of logarithms to solve the following equation:

$$2\log_3 x = \log_9 16$$

Write down the **Summary of Logarithm Properties** as seen on page 5.3-18.

Section 5.4 Guided Notebook

Section 5.4 Exponential and Logarithmic Equations

- ☐ Work through Section 5.4 TTK #1
- ☐ Work through Section 5.4 TTK #2
- ☐ Work through Section 5.4 TTK #3
- ☐ Work through Section 5.4 TTK #5
- ☐ Work through Section 5.4 TTK #6
- ☐ Work through Objective 1
- ☐ Work through Objective 2

GET TO KNOW YOUR CALCULATOR!

In this section, it is crucial that you know how to correctly use your scientific calculator. Test yourself by trying the following calculator exercises. If you have trouble getting the answers below, please seek help.

1. $e^{-0.0035} \approx .9965$

2. $\dfrac{5e^2 - 4}{e^2 - 1} \approx 5.1565$

3. $\ln 5 \approx 1.6094$

4. $\ln\left(\dfrac{7}{2}\right) \approx 1.2528$

5. $\dfrac{\ln 3 + \ln 5}{\ln 5} \approx 1.6826$

6. $\dfrac{\ln \pi - 3\ln 4}{\ln \pi - 2\ln 4} \approx 1.8516$

7. $\dfrac{\ln\left(\dfrac{17}{3}\right)}{.00235} \approx 738.1281$

8. $\dfrac{\ln\left(\dfrac{77}{131}\right)}{\ln\left(\dfrac{120}{131}\right)} \approx 6.0588$

9. $$\frac{\ln 2}{12\ln\left(1+\frac{.06}{12}\right)} \approx 11.5813$$

10. $$\frac{-\ln 65}{\left(\frac{\ln\left(\frac{37}{117}\right)}{20}\right)} \approx 72.5188$$

Section 5.4 Exponential and Logarithmic Equations

Section 5.4 Objective 1 Solving Exponential Equations

Work through the video that accompanies Example 1: Solve $2^{x+1} = 3$.

Write down the **Solving Exponential Equations** summary found in your eText here:

Work through the interactive video that accompanies Example 2:
Solve each equation. For part b, round to four decimal places.

a. $3^{x-1} = \left(\dfrac{1}{27}\right)^{2x+1}$

b. $7^{x+3} = 4^{2-x}$

Work through the interactive video that accompanies Example 3:
Solve each equation. Round to four decimal places.

a. $25e^{x-5} = 17$

b. $e^{2x-1} \cdot e^{x+4} = 11$

Section 5.4 Objective 2 Solving Logarithmic Equations

Write down the **logarithm property of equality:**

Write down the three **Properties of Logarithms** shown in Objective 2 of your eText.

Work through the video that accompanies Example 4:

Solve $2\log_5(x-1)=\log_5 64$.

Does the equation above have any extraneous solutions?

What is an extraneous solution?

Write down the steps for **Solving Logarithmic Equations**.

1.

2.

3.

4.

5.

Work through the video that accompanies Example 5 and take notes here.
(Refer to the 5 steps to solving logarithmic equations above.)

Solve $\log_4(2x-1)=2$.

Work through the interactive video that accompanies Example 6 and take notes here.
(Refer to the 5 steps to solving logarithmic equations that you wrote on the previous page.)

Solve $\log_2(x+10)+\log_2(x+6)=5$.

Work through Example 7 and take notes here.

Solve $\ln(x-4)-\ln(x-5)=2$. Round to four decimal places.

Section 5.5 Guided Notebook

Section 5.5 Applications of Exponential and Logarithmic Functions

- ☐ Work through Objective 1
- ☐ Work through Objective 2
- ☐ Work through Objective 3
- ☐ Work through Objective 4

Section 5.5 Applications of Exponential and Logarithmic Equations

Section 5.5 Objective 1 Solving Compound Interest Applications

Complete the two compound interest formulas seen here:

Compound Interest Formulas

Periodic Compound Interest Formula

$$A = \_\_\_\_\_\_\_\_\_\_\_\_$$

Continuous Compound Interest Formula

$$A = \_\_\_\_\_\_\_\_\_\_\_\_$$

where

A = Total amount after t years
P = Principal (original investment)
r = Interest rate per year
n = Number of times interest is compounded per year
t = Number of years

Work through the video that accompanies Example 1 and take notes here:

How long will it take (in years and months) for an investment to double if it earns 7.5% compounded monthly?

Work through the video that accompanies Example 2 and take notes here:

Suppose an investment of $5,000 compounded continuously grew to an amount of $5,130.50 in 6 months. Find the interest rate, and then determine how long it will take for the investment to grow to $6,000. Round the interest rate to the nearest hundredth of a percent and the time to the nearest hundredth of a year.

Section 5.5 Objective 2 Exponential Growth and Decay

Exponential Growth: When a population grows at a rate proportional to the size of the current population, the following exponential growth function is used. $P(t) = P_0 e^{kt}$ where $k > 0$ and P_0 (sometimes called "P-not") is the initial population. The graph of this exponential growth function is seen below. Note that k is a constant called the **relative growth rate.**

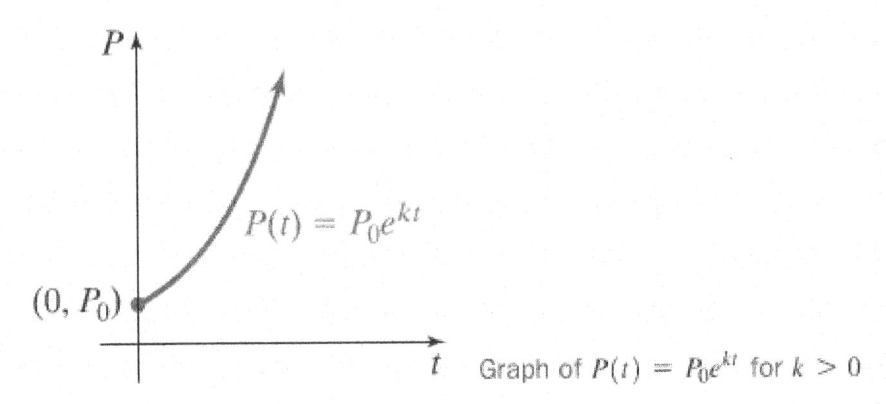

Graph of $P(t) = P_0 e^{kt}$ for $k > 0$

Work through the video that accompanies Example 3:

The population of a small town grows at a rate proportional to its current size. In 1900, the population was 900. In 1920, the population had grown to 1,600. What was the population of this town in 1950? Round to the nearest whole number.

Exponential Decay
A model that describes the exponential decay of a population, quantity, or amount A, after a certain time, t, is $A(t) = A_0e^{kt}$, where $k < 0$ and A_0 (sometimes called "A-not") is the initial quantity. The graph of this exponential decay function is seen below.

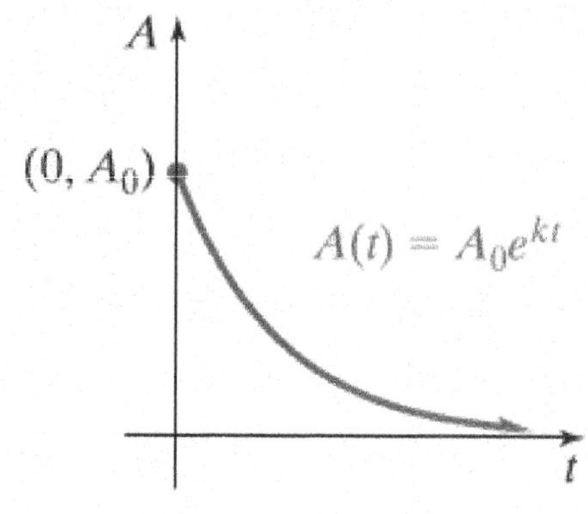

Graph of $A(t) = A_0e^{kt}$ for $k < 0$

One very useful application of the exponential decay model is **half-life.**

Work through the animation that describes half-life.

Define **half-life:**

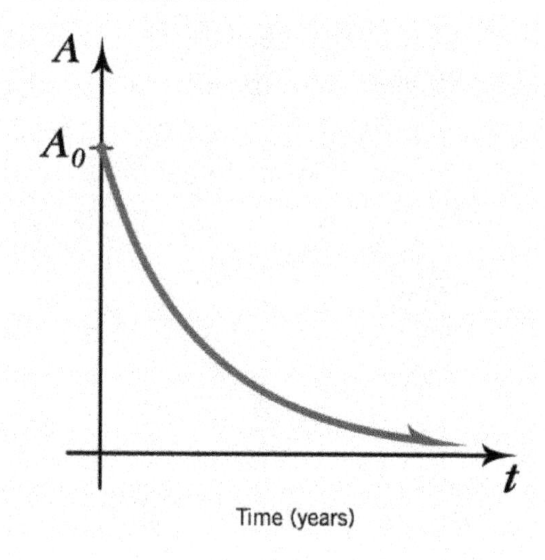

What is the half-life of cesium-137?

On the graph on the previous page, insert the relevant time values and corresponding A values as seen in this animation. How much Cesium-137 would be left after 120 years? 150 years?

Work through the video that accompanies Example 4:
Suppose that a meteorite is found containing 4% of its original krypton-99. If the half-life of krypton-99 is 80 years, how old is the meteorite? Round to the nearest year.

Section 5.5 Objective 3 Solving Logistic Growth Applications

Another population growth model is used when there are outside factors that affect the population growth. The outside factors may include predators, physical space, disease, etc. This model is called a logistic growth model. Read the introduction paragraph to Objective 3 on page 5.5-14 of your eText.

Logistic Growth

A model that describes the logistic growth of a population P at any time t is given by the function

$$P(t) = \frac{C}{1+Be^{kt}},$$

where B, C, and k are constants with $C > 0$ and $k < 0$.

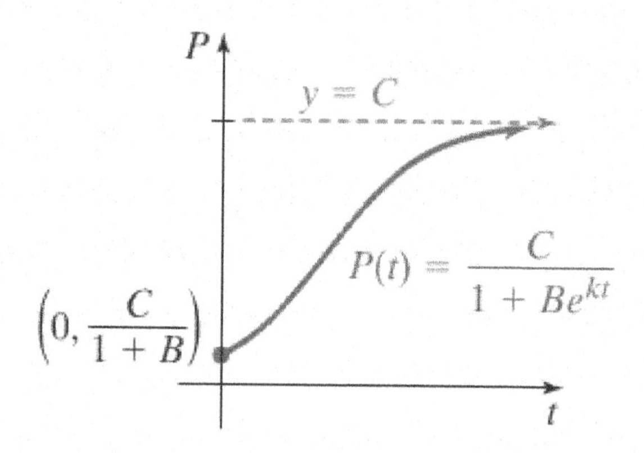

In the logistic growth model, what does the number C represent?

Work through the interactive video that accompanies Example 5:

Ten goldfish were introduced into a small pond. Because of limited food, space, and oxygen, the carrying capacity of the pond is 400 goldfish. The goldfish population at any time t, in days, is modeled by the logistic growth function $F(t) = \frac{C}{1+Be^{kt}}$. If 30 goldfish are in the pond after 20 days,

a. Find B.

b. Find k.

c. When will the pond contain 250 goldfish? Round to the nearest whole number.

Section 6.1 Guided Notebook

6.1 An Introduction to Angles: Degree and Radian Measure

- ☐ Work through Section 6.1 TTK #1
- ☐ Work through Section 6.1 Objective 1
- ☐ Work through Section 6.1 Objective 2
- ☐ Work through Section 6.1 Objective 3
- ☐ Work through Section 6.1 Objective 4
- ☐ Work through Section 6.1 Objective 5

Section 6.1 An Introduction to Angles: Degree and Radian Measure

Section 6.1 Introduction

What is the definition of a **vertex**?

What is the definition of the **initial side**?

What is the definition of the **terminal side**?

Sketch an angle with positive measure, labeling the vertex, initial side, and terminal side. Do the same for an angle with negative measure.

What does it mean for an angle to be in **standard position**?

Sketch a coordinate plane and label the **four quadrants**.

Section 6.1 Objective 1 Understanding Degree Measure

In the **degree measure** system, what is the symbol used to indicate a degree? How many degrees are in a one complete counterclockwise rotation?

Sketch three coordinate planes, illustrating angles of 360, 90, and –45 degrees, respectively. (See Figures 3, 4, and 5.)

What is the definition of an **acute angle**?

What is the definition of an **obtuse angle**?

What is the definition of a **quadrantal angle**?

What is the term for an angle of exactly 90 degrees?

What is the term for an angle of exactly 180 degrees?

What does it mean for angles to be **coterminal**?

Sketch the two coordinate planes illustrating common positive and negative angles as seen in Figure 6.

Work through the video accompanying Example 1 showing all work below.
Draw each angle in standard position and state the quadrant in which the terminal side of the angle lies or the axis on which the terminal side of the angle lies.

a. $\theta = 60°$

b. $\alpha = -270°$

c. $\beta = 420°$

Section 6.1 Objective 2 Finding Coterminal Angles Using Degree Measure

What is the definition of **Coterminal Angles**?

Starting with a given angle, how can you obtain coterminal angles?
(See the **coterminal angle** definition box.)

What notation is used to denote the angle of least nonnegative measure that is coterminal with θ?

Work through the video with Example 2 and show all work below.
Find the angle of least nonnegative measure, θ_C, that is coterminal with $\theta = -697°$.

Section 6.1 Objective 3 Understanding Radian Measure

Carefully work through the **animation** seen next to Objective 3 on page 6.1-13 and answer the questions below:

Draw a circle centered at the origin having a radius of r units. What is the equation of the circle?

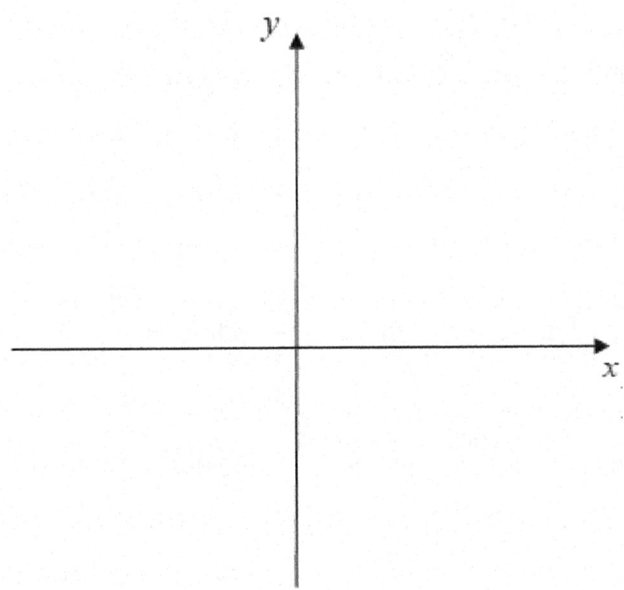

What is the definition of a **central angle**?

What is the definition of an **intercepted arc**? What variable is typically used to represent it?

On the graph of your circle above, draw a central angle so that the intercepted arc is the same length as the radius of your circle.

What is the measure of this central angle called?

What is the definition of a **radian**?

Approximately how many radians are there in a circle?

Carefully work through the **animation** seen near the bottom of page 6.1-13 and answer the questions below:

What is the formula for the circumference of a circle of radius r units?

(Fill in the blank) A central angle of $360°$ intercepts an arc length of ________________.

Complete the proportion below as seen in the animation.

$$\frac{360°}{\square} = \frac{\square}{r}$$

Fill in the box: $360° = \square$ radians.

Fill in the box: $180° = \square$ radians.

Sketch three coordinate planes, illustrating angles of 2π, $\frac{\pi}{2}$, and $\frac{-\pi}{4}$ radians, respectively.

(See Figures 10, 11, and 12.)

Sketch two coordinate planes illustrating common positive and negative angles in radians, as seen in Figure 13.

Work through the interactive video accompanying Example 3 showing all work below. Draw each angle in standard position and state the quadrant in which the terminal side of the angle lies or the axis on which the terminal side of the angle lies.

a. $\theta = \frac{\pi}{3}$

b. $\alpha = -\frac{3\pi}{2}$

c. $\beta = \frac{7\pi}{3}$

Section 6.1 Objective 4 Converting between Degree Measure and Radian Measure

To convert **degrees to radians**, multiply by ________________________.

To convert **radians to degrees**, multiply by ________________________.

Work through the interactive video with Example 4 and show all work below.
Convert each angle given in degree measure into radians.

a. 45°

b. −150°

c. 56°

Work through the interactive video with Example 5 and show all work below. Convert each angle given in radian measure into degrees. Round to two decimal places if needed.

a. $\frac{2\pi}{3}$ radians

b. $-\frac{11\pi}{6}$ radians

c. 3 radians

Section 6.1 Objective 5 Finding Coterminal Angles Using Radian Measure

For any angle θ and for any nonzero integer k, we can find a coterminal angle using what expression?

Work through Example 6 and show all work below.

Find three angles that are coterminal with $\theta = \dfrac{\pi}{3}$ using $k = 1$, $k = -1$, and $k = 2$.

Work through the video with Example 7 and show all work below.

Find the angle of least nonnegative measure, θ_C, that is coterminal with $\theta = -\frac{21\pi}{4}$.

Section 6.4 Guided Notebook

6.4 Right Triangle Trigonometry

- ☐ Work through Section 6.4 TTK #1–3
- ☐ Work through Section 6.4 Objective 1
- ☐ Work through Section 6.4 Objective 2
- ☐ Work through Section 6.4 Objective 3
- ☐ Work through Section 6.4 Objective 4
- ☐ Work through Section 6.4 Objective 5

Section 6.4 Right Triangle Trigonometry

Section 6.4 Objective 1 Understanding the Right Triangle Definitions of the Trigonometric Functions

Sketch and label the two triangles seen in Figure 29.

What are the **Right Triangle Definitions of the Trigonometric Functions**?

Work through the interactive video with Example 1 showing all work below.
Given the right triangle evaluate the six trigonometric functions of the acute angle θ.

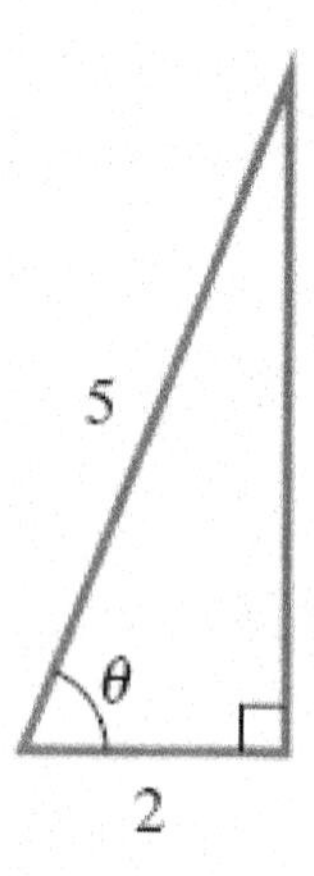

Work through the video with Example 2 showing all work below.

If θ is an acute angle of a right triangle and if $\sin\theta = \frac{3}{4}$, then find the values of the remaining five trigonometric functions for angle θ.

Section 6.4 Objective 2 Using the Special Right Triangles

Copy down the trigonometric functions for acute angles $\frac{\pi}{6}, \frac{\pi}{4}, \frac{\pi}{3}$ as seen in Table 1.

θ	$\frac{\pi}{6}(30°)$	$\frac{\pi}{4}(45°)$	$\frac{\pi}{3}(60°)$
$\sin\theta$			
$\cos\theta$			
$\tan\theta$			

Work through the video with Example 3 and show all work below.

Determine the value of $\csc\frac{\pi}{6}+\cot\frac{\pi}{4}$.

Work through Example 4 and show all work below.

Determine the measure of the acute angle θ for which $\sec\theta = 2$.

Section 6.4 Objective 3 Understanding the Fundamental Trigonometric Identities

What are the **Quotient Identities**?

What are the **Reciprocal Identities**?

Work through the video that accompanies Example 5 showing all work below.

Given that $\sin\theta = \frac{5}{7}$ and $\cos\theta = \frac{2\sqrt{6}}{7}$, find the values of the remaining four trigonometric functions using identities.

What are the **Pythagorean Identities**?

Work through the interactive video with Example 6 showing all work below.
Use identities to find the exact value of each trigonometric expression.

a. $\tan 37° - \dfrac{\sin 37°}{\cos 37°}$

b. $\dfrac{1}{\cos^2 \frac{\pi}{9}} - \dfrac{1}{\cot^2 \frac{\pi}{9}}$

Section 6.4 Objective 4 Understanding Cofunctions

What does it mean for two angles to be **complementary**?

What are the **Cofunction Identities**?

Work through the interactive video with Example 7 and show all work below.

a. Rewrite the expression $(\cot(\frac{\pi}{2}-\theta))\cos\theta$ as one of the six trigonometric functions of acute angle θ.

b. Determine the exact value of $\sec 55° \csc 35° - \tan 55° \cot 35°$.

Section 6.4 Objective 5 Evaluating Trigonometric Functions Using a Calculator

Work through the video with Example 8 and show all work below.
Evaluate each trigonometric expression using a calculator. Round each answer to four decimal places.

a. $\sin\frac{8\pi}{7}$

b. $\cot 70°$

c. $\sec\frac{\pi}{5}$

Write down the **Caution** as seen on page 6.4-29:

Section 6.5 Guided Notebook

6.5 Trigonometric Functions of General Angles

- ☐ Work through Section 6.5 TTK #1
- ☐ Work through Section 6.5 TTK #2
- ☐ Work through Section 6.5 TTK #3
- ☐ Work through Section 6.5 TTK #4
- ☐ Work through Section 6.5 TTK #5
- ☐ Work through Section 6.5 Objective 1
- ☐ Work through Section 6.5 Objective 2
- ☐ Work through Section 6.5 Objective 3
- ☐ Work through Section 6.5 Objective 4
- ☐ Work through Section 6.5 Objective 5
- ☐ Work through Section 6.5 Objective 6

Section 6.5 Trigonometric Functions of General Angles

Section 6.5 Objective 1 Understanding the Four Families of Special Angles

What is the **Quadrantal Family of Angles**? Sketch the angles shown in Figure 35.

What is the $\frac{\pi}{3}$ **Family of Angles**? Sketch the angles shown in Figure 36.

What is the $\frac{\pi}{6}$ **Family of Angles**? Sketch the angles shown in Figure 37.

What is the $\frac{\pi}{4}$ **Family of Angles**? Sketch the angles shown in Figure 38.

Work through the interactive video with Example 1 showing all work below.
Each of the given angles belongs to one of the four families of special angles. Determine the family of angles to which it belongs, sketch the angle, and then determine the angle of least nonnegative measure, θ_C, coterminal with the given angle.

a. $\theta = \frac{29\pi}{6}$

b. $\theta = \frac{14\pi}{2}$

c. $\theta = -\frac{18\pi}{4}$

d. $\theta = \frac{11\pi}{4}$

e. $\theta = \frac{14\pi}{6}$

f. $\theta = 420°$

g. $\theta = -495°$

Section 6.5 Objective 2 Understanding the Definitions of the Trigonometric Functions of General Angles

What are the **General Angle Definitions of the Trigonometric Functions**?

Under what conditions will the following trigonometric functions be undefined (if ever)?

$\tan\theta = \frac{y}{x}$ and $\sec\theta = \frac{r}{x}$:

$\csc\theta = \frac{r}{y}$ and $\cot\theta = \frac{x}{y}$:

$\sin\theta = \frac{y}{r}$ and $\cos\theta = \frac{x}{r}$:

Work through the video with Example 2 and show all work below.
Suppose that the point $(-4,-6)$ is on the terminal side of an angle θ. Find the six trigonometric functions of θ.

Section 6.5 Objective 3 Finding the Values of the Trigonometric Functions of Quadrantal Angles

The value of the six trigonometric functions of a quadrantal angle is 0, 1, -1, or undefined. Fill in the table below. This table represents the values of the six trigonometric functions of quadtrantal angles.

θ	$\sin\theta$	$\cos\theta$	$\tan\theta$	$\csc\theta$	$\sec\theta$	$\cot\theta$
0						
$\frac{\pi}{2}$						
π						
$\frac{3\pi}{2}$						

Work through the interactive video accompanying Example 3 showing all work below. Without using a calculator, determine the value of the trigonometric function or state that the value is undefined.

a. $\cos(-11\pi)$

b. $\csc(-270°)$

c. $\tan\left(\frac{13\pi}{2}\right)$

d. $\sin(540°)$

e. $\cot\left(-\frac{7\pi}{2}\right)$

Section 6.5 Objective 4 Understanding the Signs of the Trigonometric Functions

The sign of each trigonometric function is determined by the ______________ in which the terminal side of the angle lies.

Which trigonometric functions are positive for all angles with a terminal side lying in the following quadrants?

Quadrant I:

Quadrant II:

Quadrant III:

Quadrant IV:

What acronym can help us remember the signs of the trigonometric functions for angles whose terminal side lies in one of the four quadrants?

Sketch the diagram shown in Figure 49.

Three grids are shown below. (See Figure 50.) The first grid represents the sign of the values of $y = \sin x$ in each quadrant. The middle grid represents the sign of the values of $y = \cos x$. The third quadrant represents the sign of the values of $y = \tan x$ in each quadrant. Place a "+" or "–" in each quadrant of each grid to represent the appropriate sign.

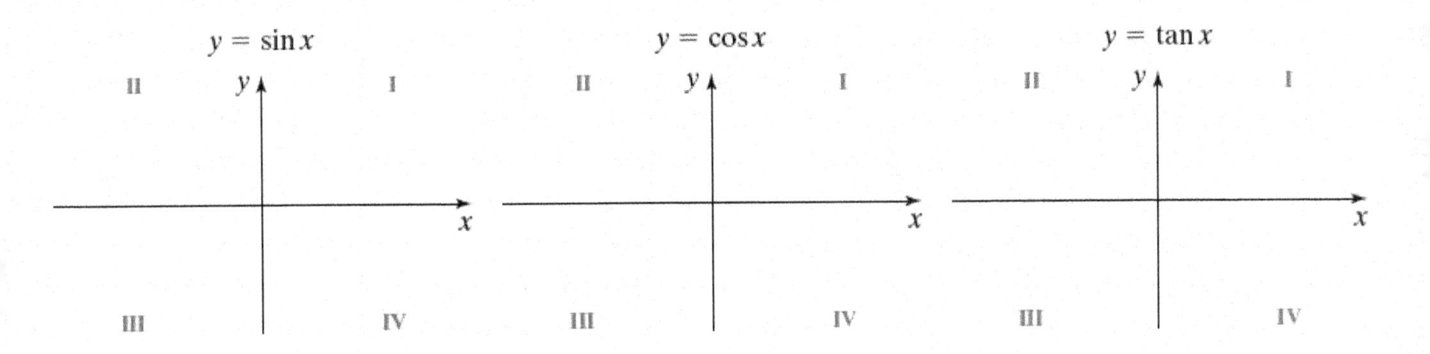

Work through the video with Example 4 and show all work below.
Suppose θ is a positive angle in standard position such that $\sin\theta < 0$ and $\sec\theta > 0$.

a. Determine the quadrant in which the terminal side of angle θ lies.

b. Find the value of $\tan\theta$ if $\sec\theta = \sqrt{5}$.

Section 6.5 Objective 5 Determining Reference Angles

What is the definition of the **Reference Angle**?

The measure of the ___________________ θ_R depends on the quadrant in

which the ______________________ of θ_C lies.

The four cases for reference angles are shown below. Fill in the blanks.

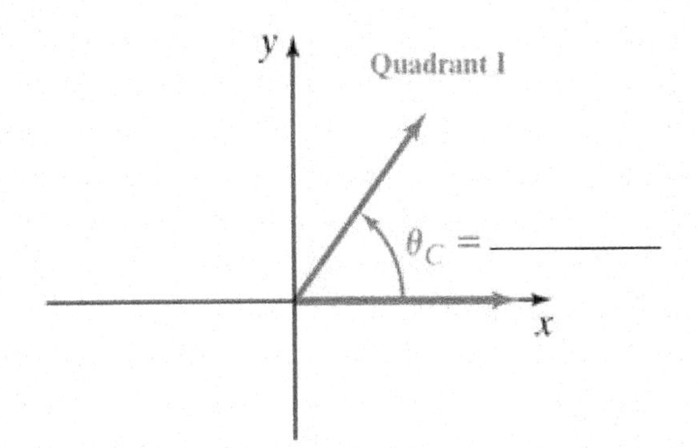

Case 1: If the terminal side of θ_C lies in Quadrant I, then θ_R = ________ .

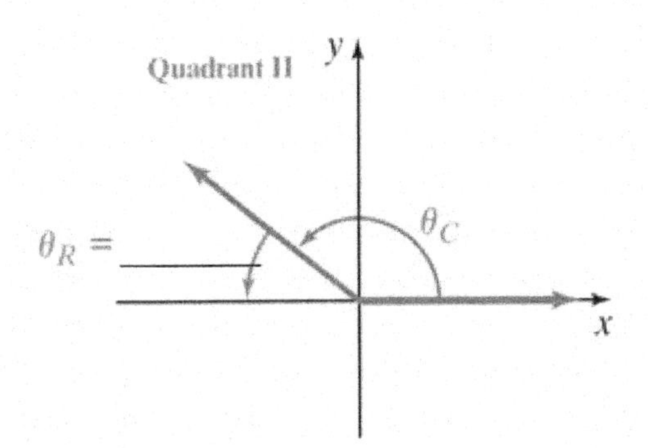

Case 2: If the terminal side of θ_C lies in Quadrant II, then θ_R = ________ (or θ_R = ________).

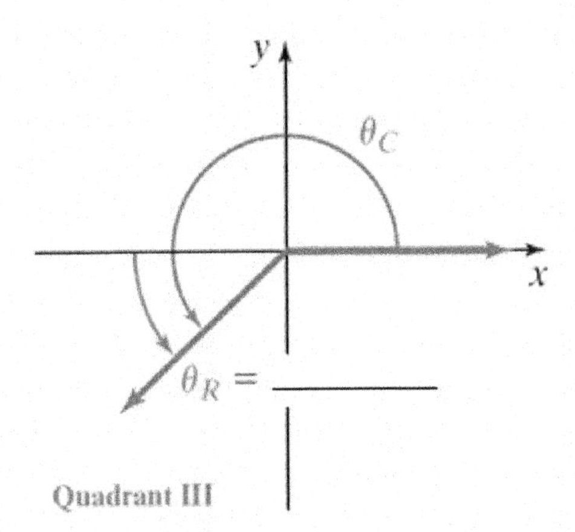

Case 3: If the terminal side of θ_C lies in Quadrant III, then θ_R = ______ (or θ_R = ________).

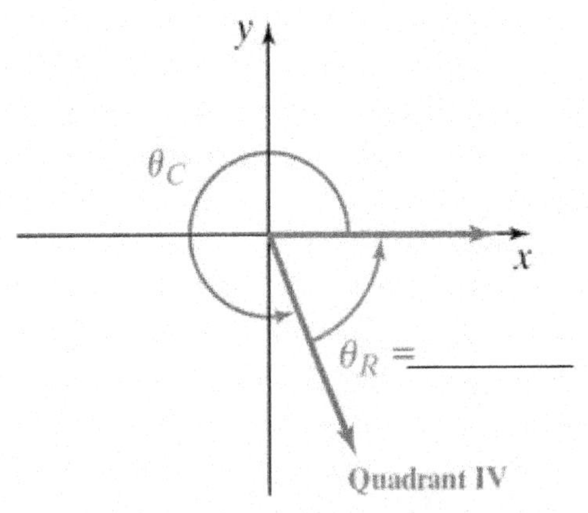

Case 4: If the terminal side of θ_C lies in Quadrant IV, then θ_R = ________ (or θ_R = ________).

Work through the interactive video with Example 5 and show all work below.
For each of the given angles, determine the reference angle.

a. $\theta = \dfrac{5\pi}{3}$

b. $\theta = \dfrac{11\pi}{4}$

c. $\theta = -\dfrac{25\pi}{6}$

d. $\theta = \dfrac{16\pi}{6}$

Work through the interactive video with Example 6 and show all work below.
For each of the given angles, determine the reference angle.

a. $\theta = \dfrac{5\pi}{8}$

b. $\theta = \dfrac{22\pi}{9}$

c. $\theta = -\dfrac{5\pi}{7}$

Work through the interactive video with Example 7 and show all work below.
For each of the given angles, determine the reference angle.

a. $\theta = 225°$

b. $\theta = -233°$

c. $\theta = 510°$

Section 6.5 Objective 6 Evaluating Trigonometric Functions of Angles Belonging to the $\pi/3$, $\pi/6$, or $\pi/4$ Families

Work through the interactive video with Example 8 and show all work below.

Find the values of the six trigonometric functions for $\theta = \frac{7\pi}{4}$.

What are the four **Steps for Evaluating Trigonometric Functions of Angles Belonging to the** $\frac{\pi}{3}$, $\frac{\pi}{6}$, **or** $\frac{\pi}{4}$ **Families**?

Step 1:

Step 2:

Step 3:

Step 4:

Work through the interactive video with Example 9 and show all work below.
Find the exact value of each trigonometric expression without using a calculator.

a. $\sin\left(\frac{7\pi}{6}\right)$

b. $\cot\left(-\frac{22\pi}{3}\right)$

c. $\tan\left(\frac{11\pi}{4}\right)$

d. $\cos\left(\frac{11\pi}{3}\right)$

e. $\sec\left(\frac{5\pi}{6}\right)$

f. $\csc\left(-\frac{7\pi}{6}\right)$

Section 6.6 Guided Notebook

6.6 The Unit Circle

- ☐ Work through Section 6.6 TTK #1-6
- ☐ Work through Section 6.6 Objective 1
- ☐ Work through Section 6.6 Objective 2
- ☐ Work through Section 6.6 Objective 3
- ☐ Work through Section 6.6 Objective 4
- ☐ Work through Section 6.6 Objective 5

Section 6.6 The Unit Circle

Section 6.6 Objective 1 Understanding the Definition of the Unit Circle

What is the **Unit Circle**? (Define it in words and sketch it.)

Work through the interactive video with Example 1 showing all work below. Determine the missing coordinate of a point that lies on the graph of the unit circle given the quadrant in which the point is located.

a. $(-\frac{1}{8}, y)$; Quadrant III

b. $(x, -\frac{\sqrt{3}}{2})$; Quadrant IV

c. $(-\frac{1}{\sqrt{2}}, y)$; Quadrant II

Section 6.6 Objective 2 Using Symmetry to Determine Points That Lie on the Unit Circle

What are the three ways in which the unit circle is symmetric?

Work through the video with Example 2 and show all work below.

Verify that the point $(-\frac{1}{8}, -\frac{3\sqrt{7}}{8})$ lies on the graph of the unit circle. Then use symmetry to find three other points that also lie on the graph of the circle.

Section 6.6 Objective 3 Understanding the Unit Circle Definitions of the Trigonometric Functions

The ________________________ of a sector of the unit circle is exactly equal

to the measure of the ________________________.

What are **The Unit Circle Definitions of the Trigonometric Functions**?
(Include the six equations and sketch the graph.)

Work through the video with Example 3 showing all work below.

If $\left(-\frac{1}{4}, \frac{\sqrt{15}}{4}\right)$ is a point on the unit circle corresponding to a real number t, find the values of the six trigonometric functions of t.

Section 6.6 Objective 4 Using the Unit Circle to Evaluate Trigonometric Functions at Increments of $\pi/2$

Use the unit circle definitions of the trigonometric functions and the figure below to fill out the table.

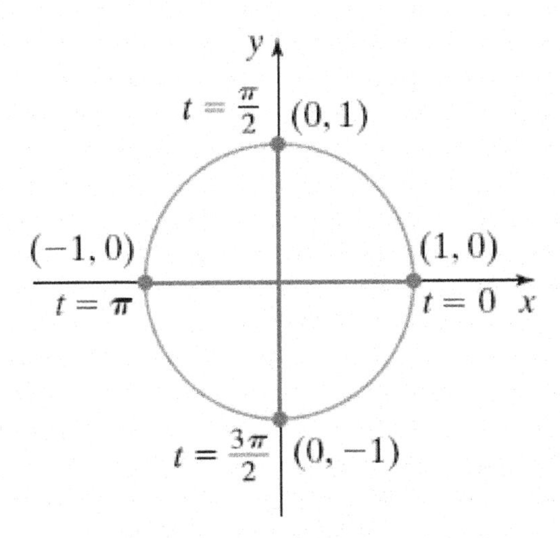

t	$\sin t$	$\cos t$	$\tan t$	$\csc t$	$\sec t$	$\cot t$
$0, 2\pi$						
$\frac{\pi}{2}$						
π						
$\frac{3\pi}{2}$						

Work through the interactive video with Example 4 and show all work below.
Use the unit circle to determine the value of each expression or state that it is undefined.

a. $\cos 11\pi$

b. $\tan \frac{5\pi}{2}$

c. $\csc\left(-\frac{3\pi}{2}\right)$

Section 6.6 Objective 5 Using the Unit Circle to Evaluate Trigonometric Functions for Increments of $\pi/6$, $\pi/4$, and $\pi/3$

Sketch and label the unit circle as seen in Figure 70.

Work through the animation with Example 5 and show all work below. Use the unit circle to determine the following values.

a. $\tan\left(\frac{7\pi}{3}\right)$

b. $\sin\left(-\frac{3\pi}{4}\right)$

c. $\sec(480°)$

d. $\csc\left(-\frac{13\pi}{3}\right)$

Carefully work through the **Guided Visualization** seen on page 6.6-23.
Choose two values of t (at least one negative). Sketch a unit circle, label the point that lies on the unit circle for each value of t, and find the values of the six trigonometric functions for each of your values of t. Below is an example:

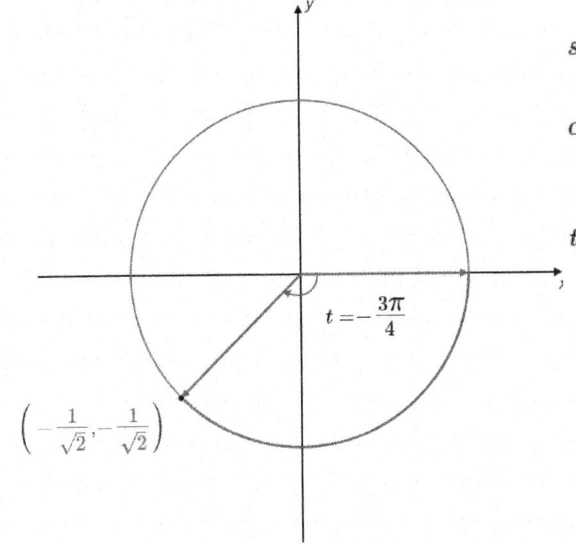

$$\sin t = y = -\frac{1}{\sqrt{2}} \qquad \csc t = \frac{1}{y} = \frac{1}{-\frac{1}{\sqrt{2}}} = -\sqrt{2}$$

$$\cos t = x = -\frac{1}{\sqrt{2}} \qquad \sec t = \frac{1}{x} = \frac{1}{-\frac{1}{\sqrt{2}}} = -\sqrt{2}$$

$$\tan t = \frac{y}{x} = \frac{-\frac{1}{\sqrt{2}}}{-\frac{1}{\sqrt{2}}} = 1 \qquad \cot t = \frac{x}{y} = \frac{-\frac{1}{\sqrt{2}}}{-\frac{1}{\sqrt{2}}} = 1$$

Section 7.1 Guided Notebook

7.1 The Graphs of Sine and Cosine

- ☐ Work through Section 7.1 TTK #1-6
- ☐ Work through Section 7.1 Objective 1
- ☐ Work through Section 7.1 Objective 2
- ☐ Work through Section 7.1 Objective 3
- ☐ Work through Section 7.1 Objective 4
- ☐ Work through Section 7.1 Objective 5
- ☐ Work through Section 7.1 Objective 6

Section 7.1 The Graphs of Sine and Cosine

Section 7.1 Objective 1 Understanding the Graph of the Sine Function and Its Properties

Watch the video that accompanies Objective 1 and fill in the tables below.

x	0	$\frac{\pi}{2}$	π	$\frac{3\pi}{2}$	2π
$y = \sin x$					

x	$\frac{\pi}{6}$	$\frac{5\pi}{6}$	$\frac{7\pi}{6}$	$\frac{11\pi}{6}$
$y = \sin x$				

x	$\frac{\pi}{4}$	$\frac{3\pi}{4}$	$\frac{5\pi}{4}$	$\frac{7\pi}{4}$
$y = \sin x$				

x	$\frac{\pi}{3}$	$\frac{2\pi}{3}$	$\frac{4\pi}{3}$	$\frac{5\pi}{3}$
$y = \sin x$				

Sketch the graph of $y = \sin x$ on the interval $[0, 2\pi]$.

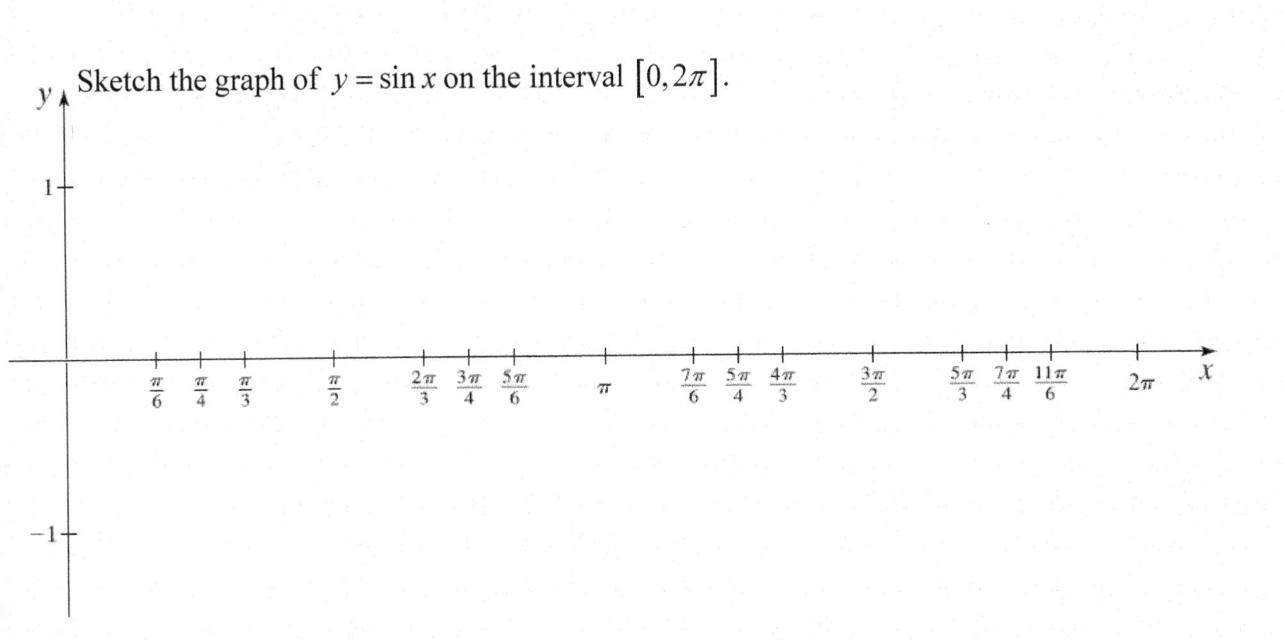

What is the definition of a **Periodic Function**?

Write down the **Characteristics of the Sine Function.**

Work through the video that accompanies Example 1 showing all work below.
Using the graph of $y = \sin x$, list all values of x on the interval $\left[-3\pi, \dfrac{7\pi}{4}\right]$ that satisfy the ordered pair $(x, 0)$.

Work through the video that accompanies Example 2 showing all work below.
Use the periodic property of $y = \sin x$ to determine which of the following expressions is equivalent to $\sin\left(\dfrac{23\pi}{6}\right)$.

i. $\sin\left(\dfrac{\pi}{6}\right)$ ii. $\sin\left(\dfrac{5\pi}{6}\right)$ iii. $\sin\left(\dfrac{11\pi}{6}\right)$ iv. $\sin\left(\dfrac{13\pi}{6}\right)$

Work through Example 3 showing all work below.
Use the fact that $y = \sin x$ is an odd function to determine which of the following expressions is equivalent to $-\sin\left(\dfrac{9\pi}{16}\right)$.

i. $\sin\left(-\dfrac{9\pi}{16}\right)$ ii. $\sin\left(\dfrac{9\pi}{16}\right)$ iii. $-\sin\left(-\dfrac{9\pi}{16}\right)$

Section 7.1 Objective 2 Understanding the Graph of the Cosine Function and Its Properties

Fill in the four tables below.

x	0	$\frac{\pi}{2}$	π	$\frac{3\pi}{2}$	2π
$y = \cos x$					

x	$\frac{\pi}{6}$	$\frac{5\pi}{6}$	$\frac{7\pi}{6}$	$\frac{11\pi}{6}$
$y = \cos x$				

x	$\frac{\pi}{4}$	$\frac{3\pi}{4}$	$\frac{5\pi}{4}$	$\frac{7\pi}{4}$
$y = \cos x$				

x	$\frac{\pi}{3}$	$\frac{2\pi}{3}$	$\frac{4\pi}{3}$	$\frac{5\pi}{3}$
$y = \cos x$				

Sketch the graph of $y = \cos x$ on the interval $[0, 2\pi]$.

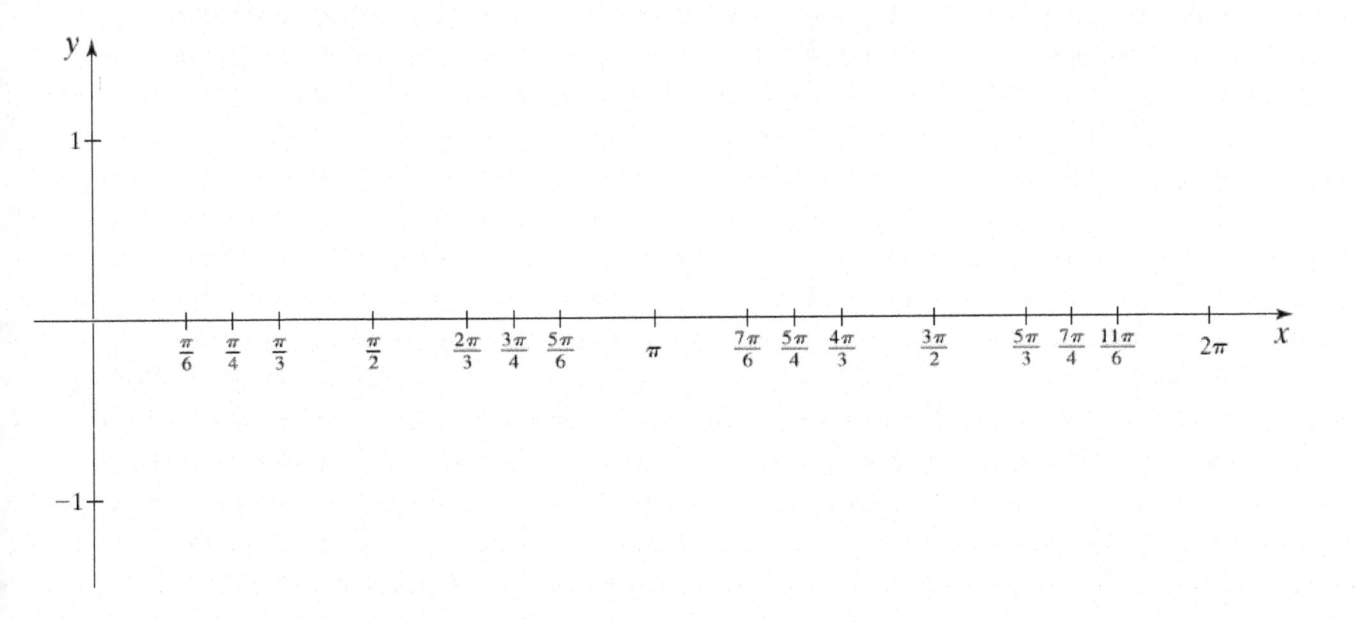

Write down the **Characteristics of the Cosine Function.**

Work through the video that accompanies Example 4 showing all work below.
Using the graph of $y = \cos x$, list all values of x on the interval $[-\pi, 2\pi]$ that satisfy the ordered pair $\left(x, \frac{1}{2}\right)$.

Write down **The Five Quarter Points of** $y = \sin x$ **and** $y = \cos x$ and plot them on the graphs below. **You must memorize these quarter points!**

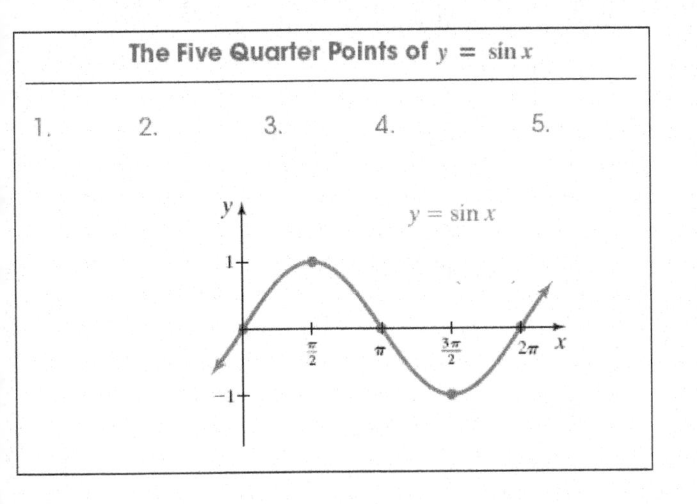

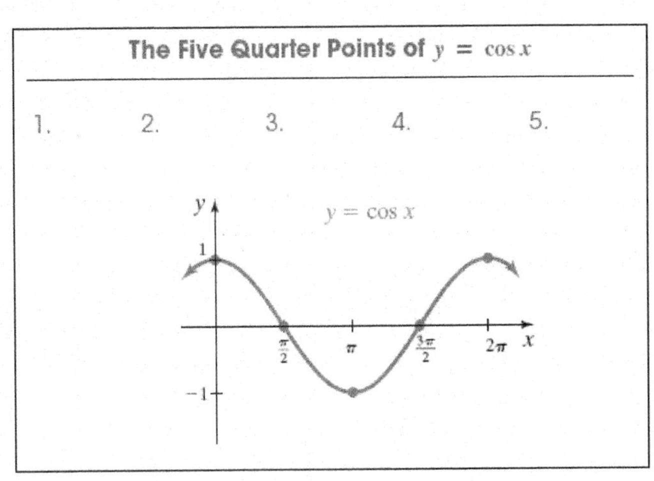

Section 7.1 Objective 3 Sketching Graphs of the From $y = A\sin x$ and $y = A\cos x$

What is the definition of **amplitude**?

Work through the video with Example 5 showing all work below.

Determine the amplitude and range of $y = -\frac{2}{3}\cos x$ and then sketch the graph.

Section 7.1 Objective 4 Sketching Graphs of the Form $y = \sin(Bx)$ and $y = \cos(Bx)$

How do you **Determine the Period of** $y = \sin(Bx)$ **and** $y = \cos(Bx)$?

Work through the interactive video with Example 6 and show all work below.
Determine the period and sketch the graph of each function.

a. $y = \sin(2x)$

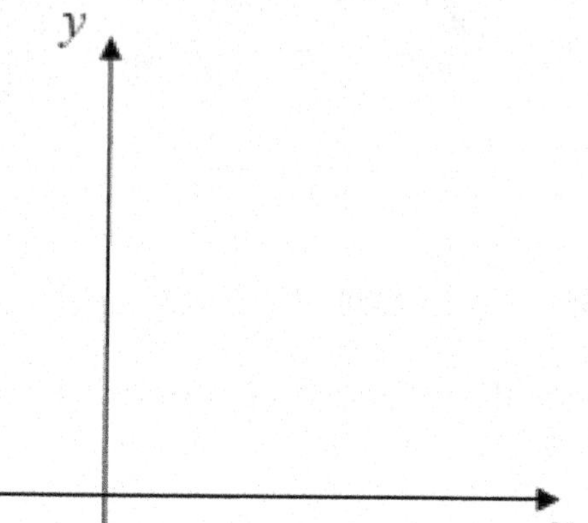

b. $y = \cos\left(\frac{1}{2}x\right)$

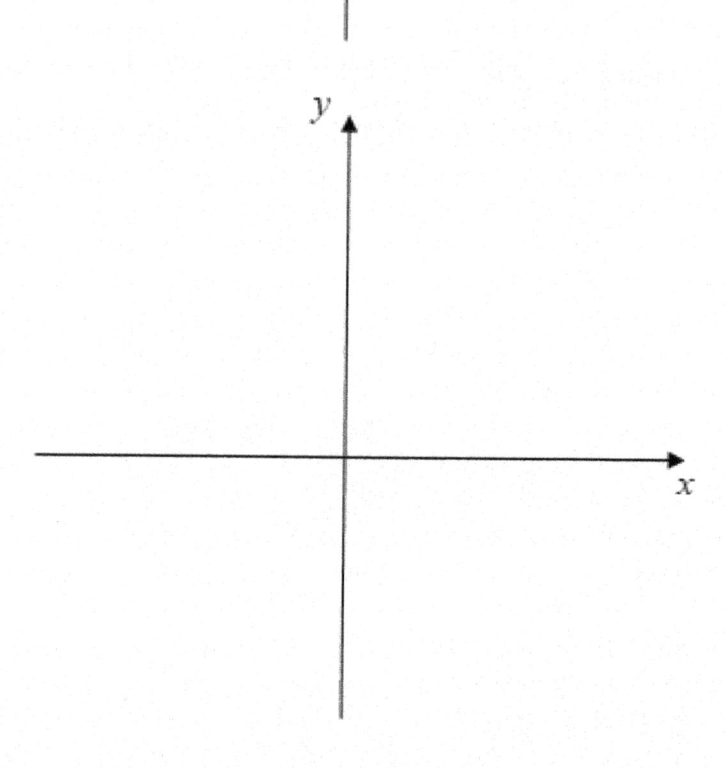

c. $y = \sin(\pi x)$

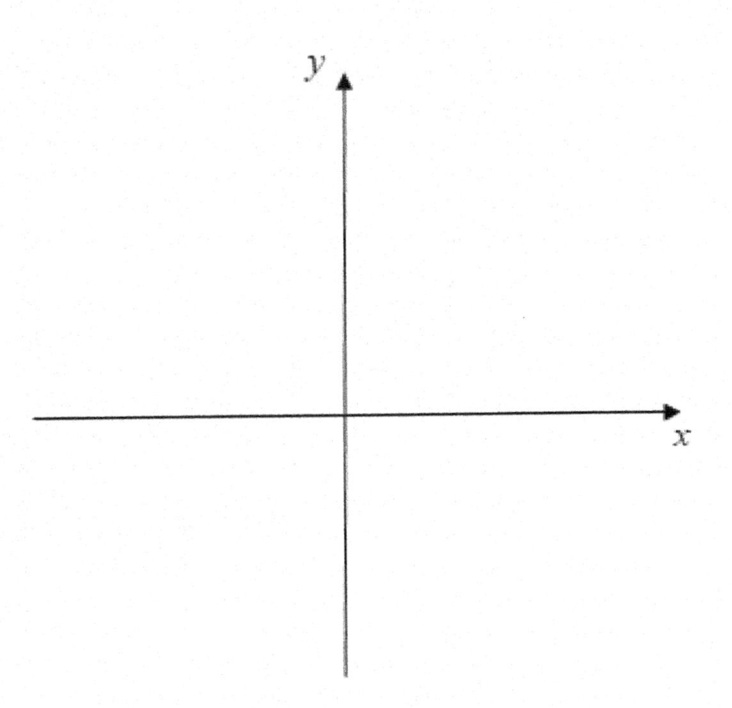

Turn to page 7.1-33 and fill in the blanks below:

Recall that $y = \sin x$ is an __________ function. This means that $\sin(-x) =$ __________ for all x in the domain of $y = \sin x$. It follows that $\sin(-Bx) =$ ______________ .

Recall that $y = \cos x$ is an __________ function. This means that $\cos(-x) =$ __________ for all x in the domain of $y = \cos x$. It follows that $\cos(-Bx) =$ ______________ .

Therefore, when $B > 0$, we can sketch the graph of $y = \sin(-Bx)$ by simply sketching the graph of __________________________ .

The graph of $y = \cos(-Bx)$ is the ___________________________ graph as the graph of ________________.

Work through the interactive video with Example 7 and show all work below.
Determine the period and sketch the graph of each function.

a. $y = \sin(-2x)$

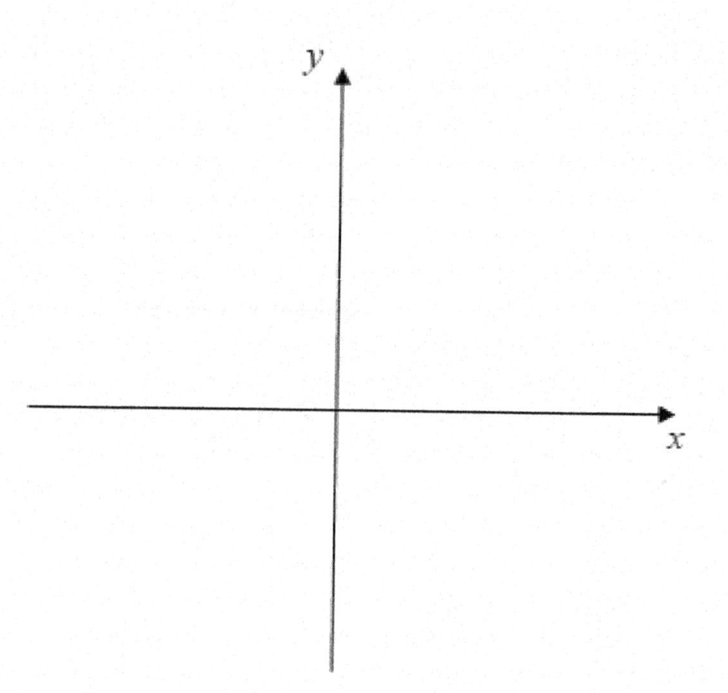

b. $y = \cos\left(-\frac{1}{2}x\right)$

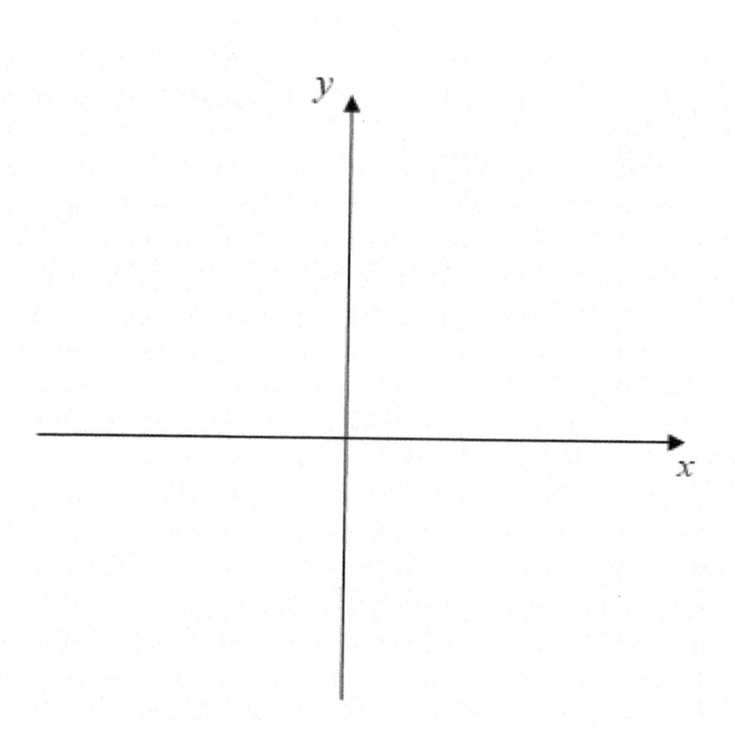

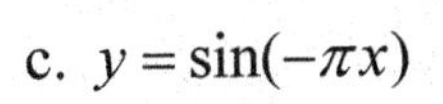

c. $y = \sin(-\pi x)$

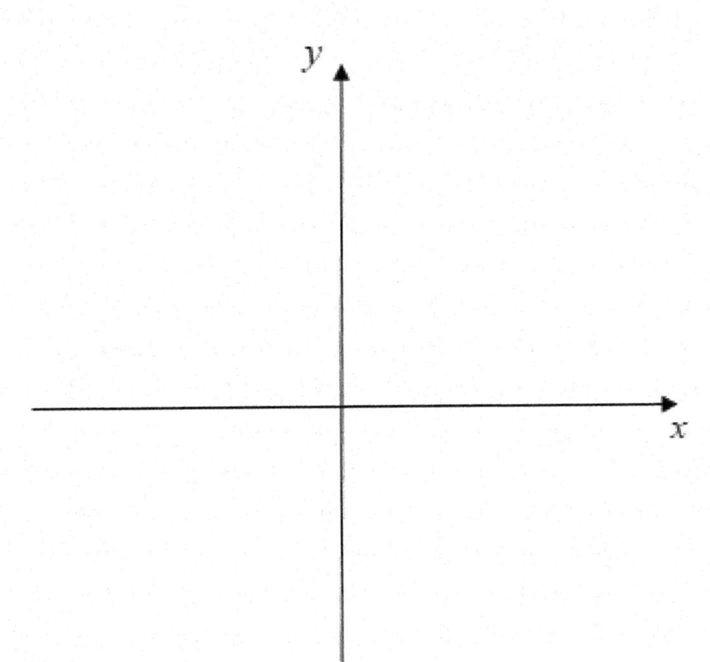

Section 7.1 Objective 5 Sketching Graphs of the Form $y = A\sin(Bx)$ and $y = A\cos(Bx)$

What are the six **Steps for Sketching Functions of the Form $y = A\sin(Bx)$ and $y = A\cos(Bx)$**?

Step 1.

Step 2.

Step 3.

Step 4.

Step 5.

Step 6.

Work through the interactive video with Example 8 and show all work below.
Use the six-step process outlined in this section to sketch each graph.

a. $y = 3\sin(4x)$

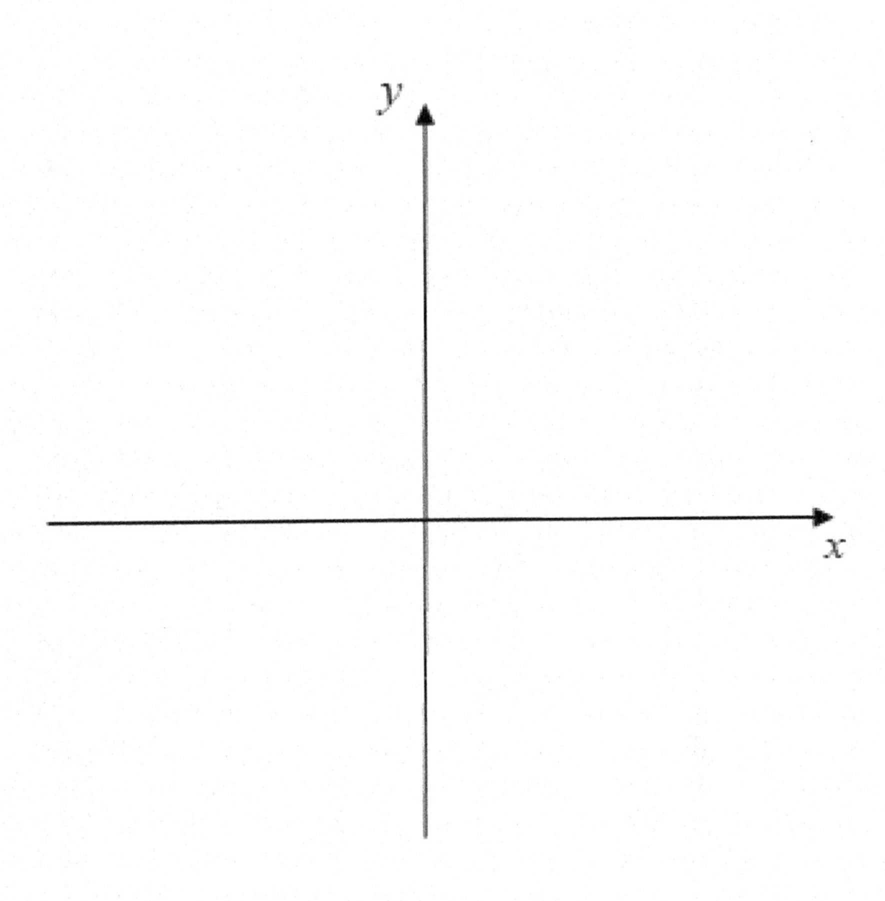

b. $y = -2\cos\left(\frac{1}{3}x\right)$

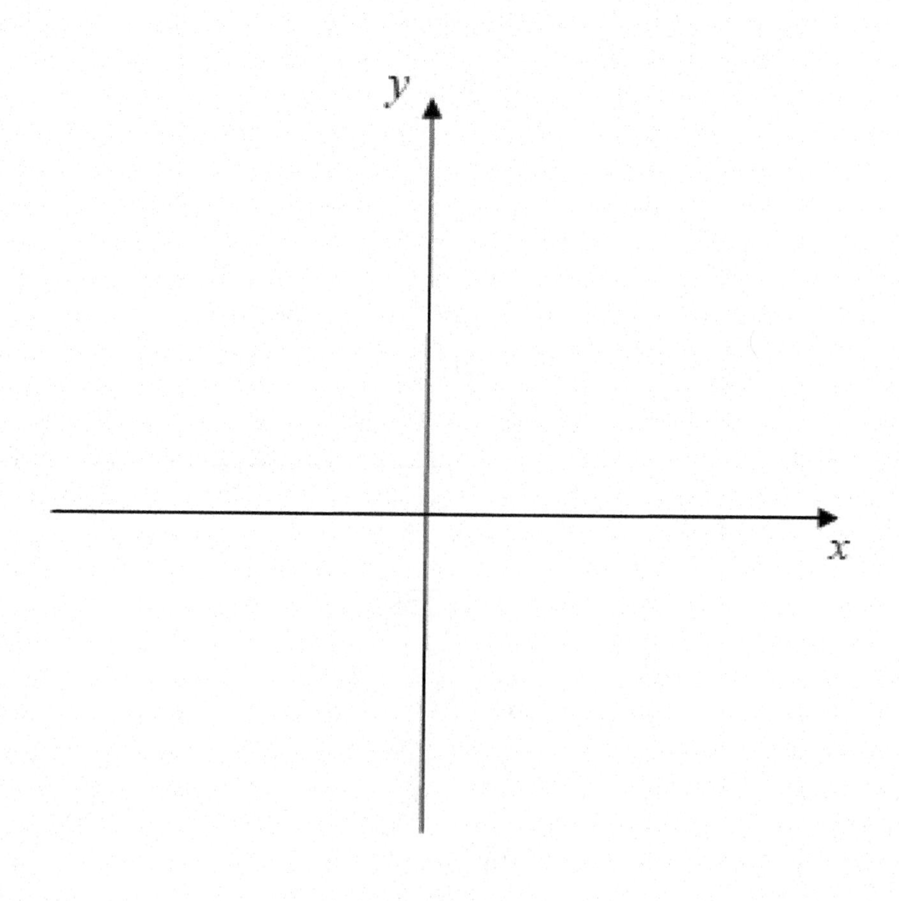

c. $y = -6\sin\left(-\dfrac{\pi x}{2}\right)$

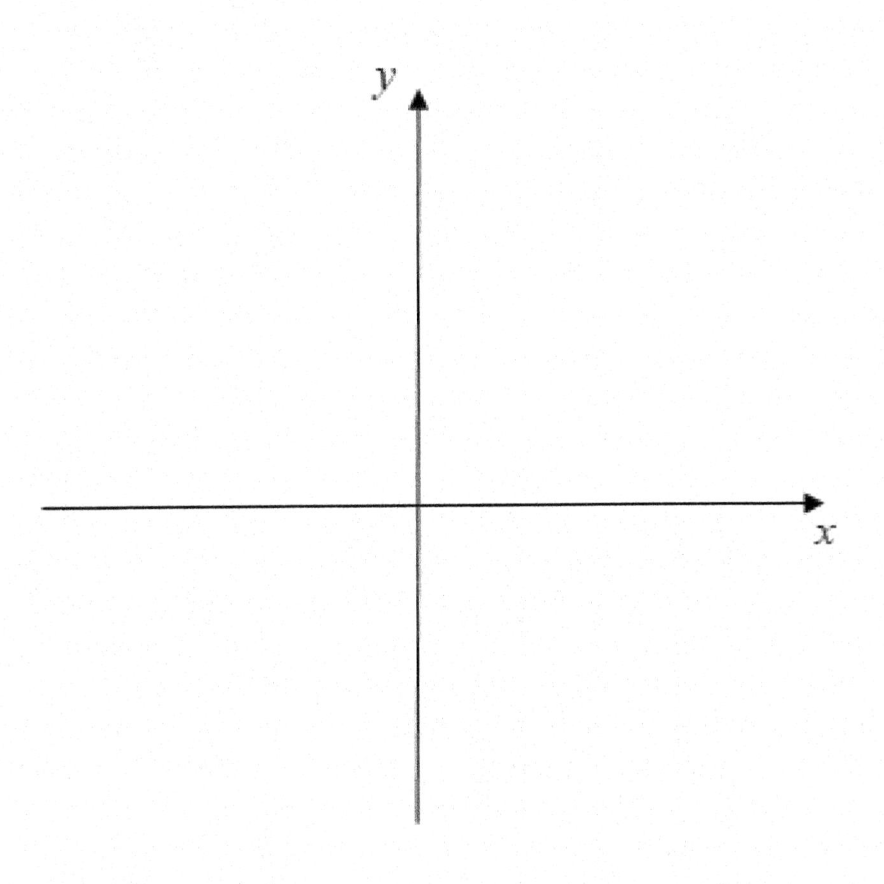

Work through the **Guided Visualization** seen on page 7.1-46. Sketch and label the graphs of $y = A\sin(Bx)$ and $y = A\cos(Bx)$. Make sure that at least one of your graphs has $A < 0$.

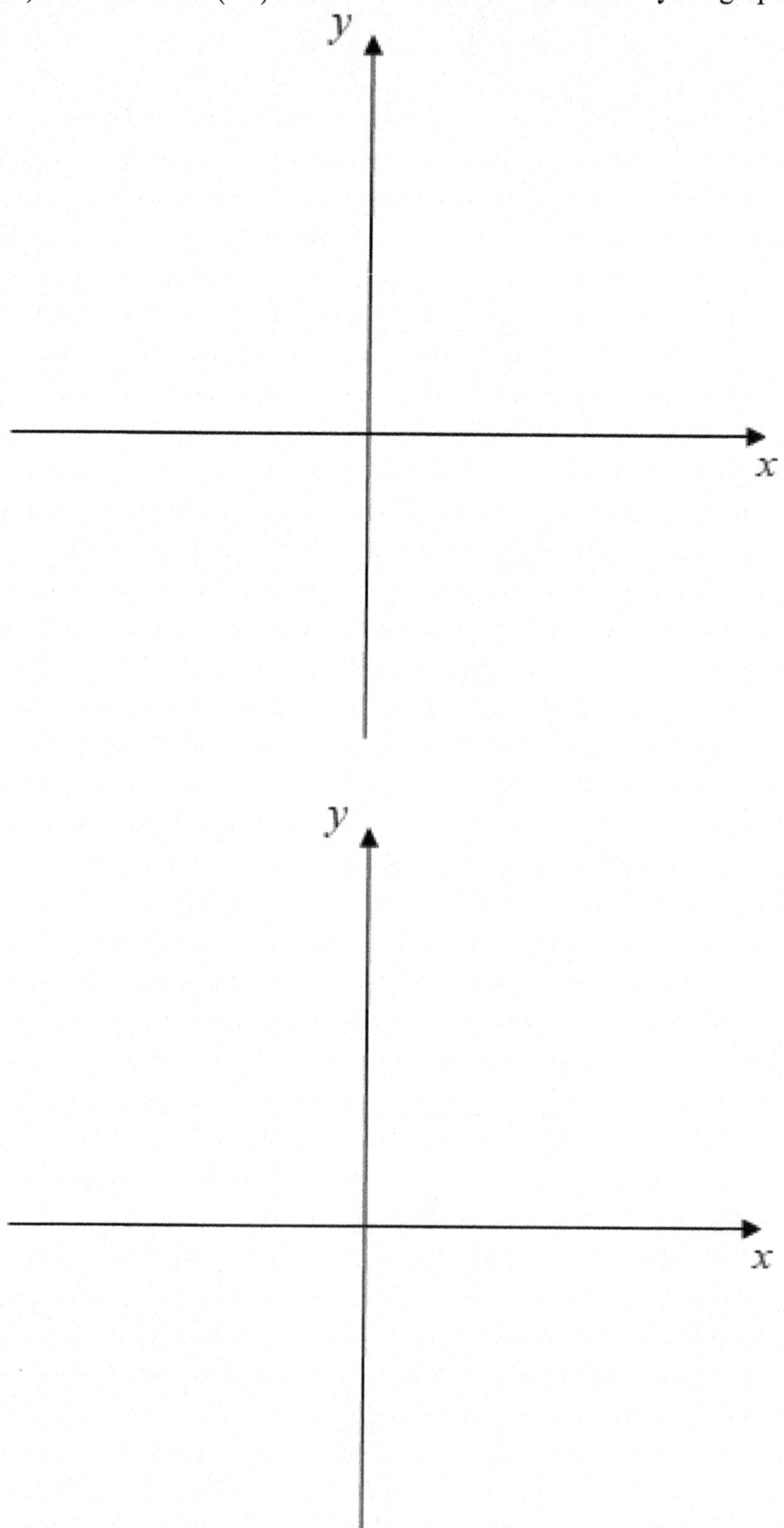

Section 7.1 Objective 6 Determine the Equation of a Function of the Form $y = A\sin(Bx)$ and $y = A\cos(Bx)$ Given its Graph

Given the graph of a function of the form $y = A\sin(Bx)$ or $y = A\cos(Bx)$, what three things must we determine in order to determine the proper function?

1.

2.

3.

Write the three steps for Determining the Equation of a Function of the form $y = A\sin(Bx)$ or $y = A\cos(Bx)$ Given the Graph.

Step 1.

Step 2.

Step 3.

Work through the interactive video that accompanies Example 9.
One cycle of the graphs of three trigonometric functions of the form $y = A\sin(Bx)$ or $y = A\cos(Bx)$ for $B > 0$ are given below. Determine the equation of the function represented by each graph.

a.

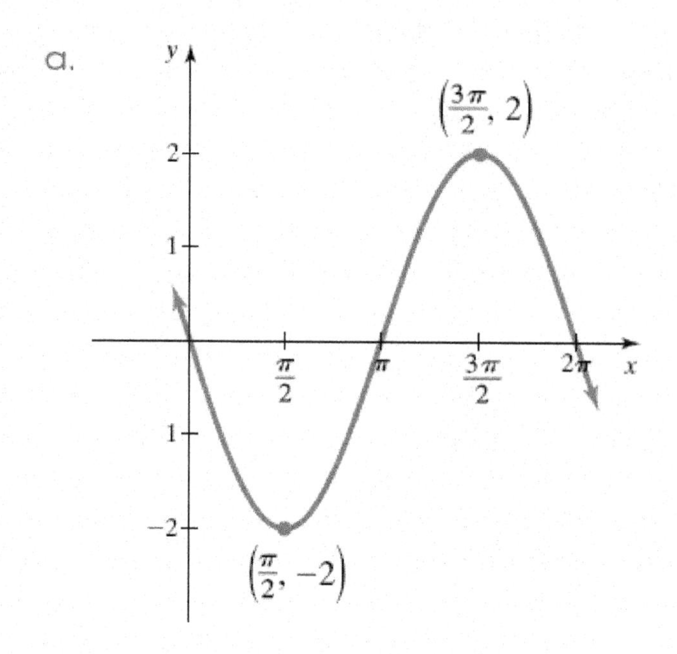

b.

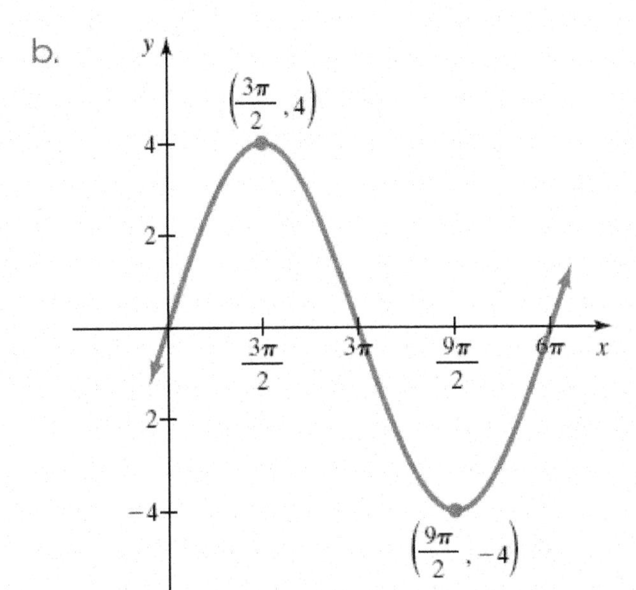

y
$\left(\frac{3\pi}{2}, 4\right)$
4
2
$\frac{3\pi}{2}$
3π
$\frac{9\pi}{2}$
6π
x
2
−4
$\left(\frac{9\pi}{2}, -4\right)$

c.

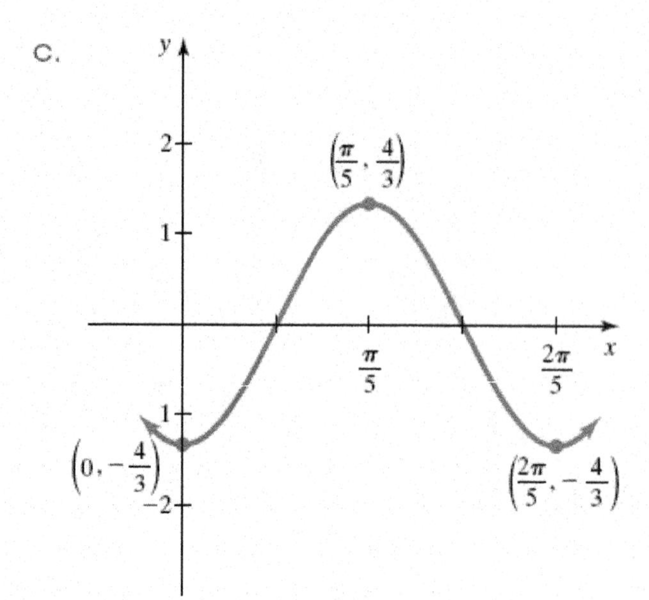
y
2
1
1
−2
x
$\frac{\pi}{5}$
$\frac{2\pi}{5}$
$\left(\frac{\pi}{5}, \frac{4}{3}\right)$
$\left(0, -\frac{4}{3}\right)$
$\left(\frac{2\pi}{5}, -\frac{4}{3}\right)$

Section 7.2 Guided Notebook

7.2 More on Graphs of Sine and Cosine: Phase Shift

- ☐ Work through Section 7.2 TTK #1–5
- ☐ Work through Section 7.2 Objective 1
- ☐ Work through Section 7.2 Objective 2
- ☐ Work through Section 7.2 Objective 3

Section 7.2 More on Graphs of Sine and Cosine: Phase Shift

Section 7.2 Objective 1 Sketching Graphs of the Form $y = \sin(x - C)$ and $y = \cos(x - C)$

Use the graphs of $y = \sin x$ and $y = \cos x$ on the left, to sketch the graphs of $y = \sin(x - C)$, $y = \sin(x + C)$, $y = \cos(x - C)$, and $y = \cos(x + C)$ for $C > 0$.

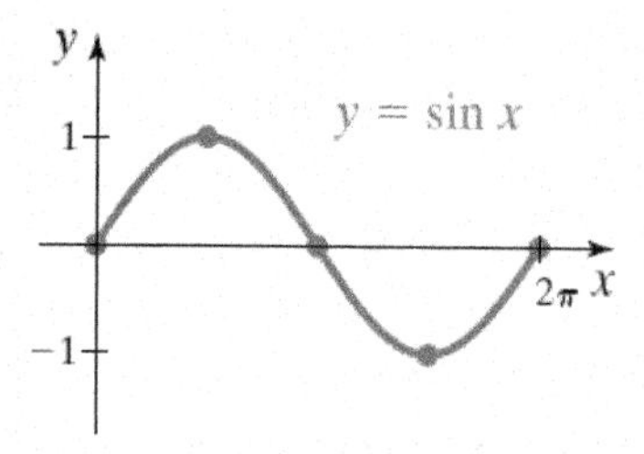

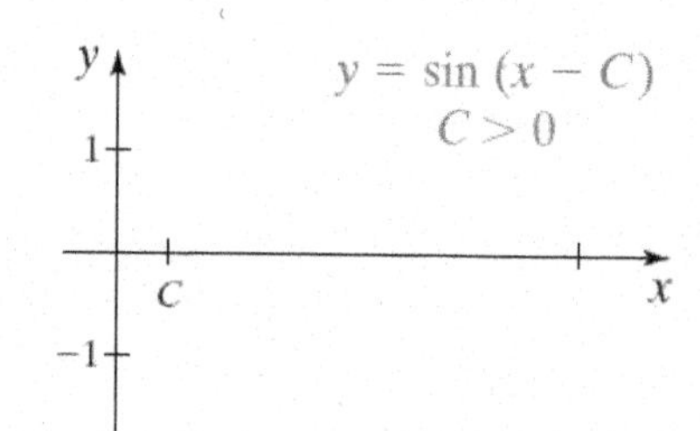

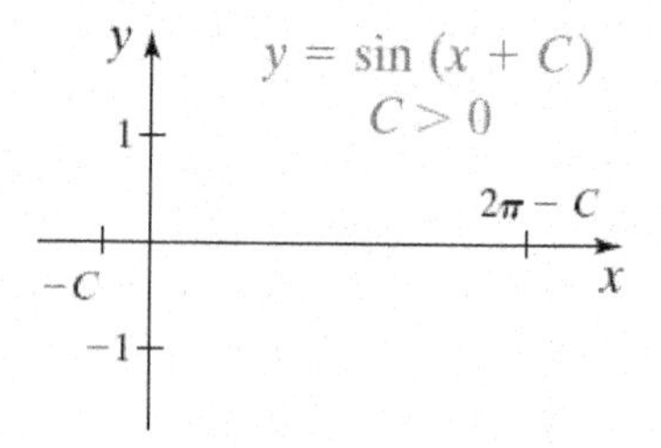

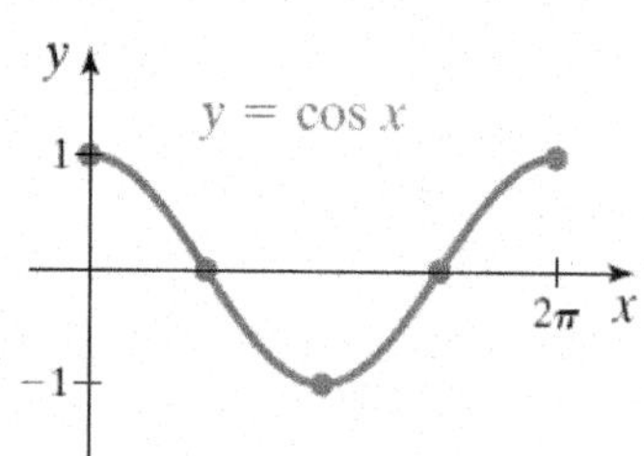

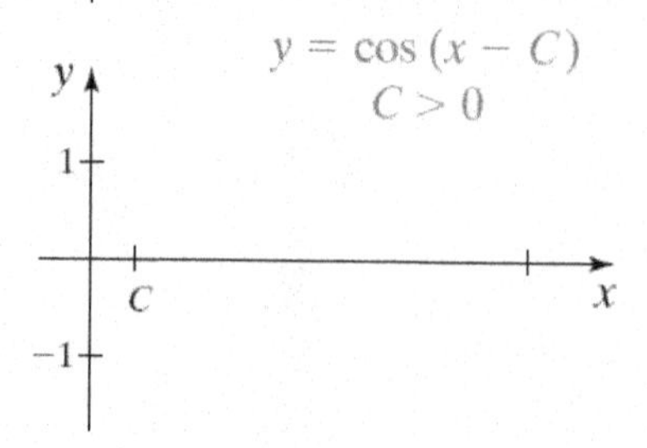

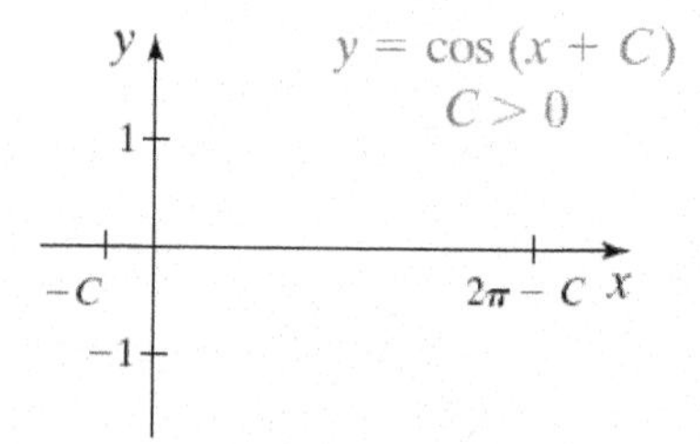

What is the **phase shift**?

Write down the properties and sketch the graphs of $y = \sin(x - C)$ and $y = \cos(x - C)$.

Work through the interactive video with Example 1 showing all work below.
Determine the phase shift and sketch the graph of each function.

a. $y = \cos(x - \pi)$

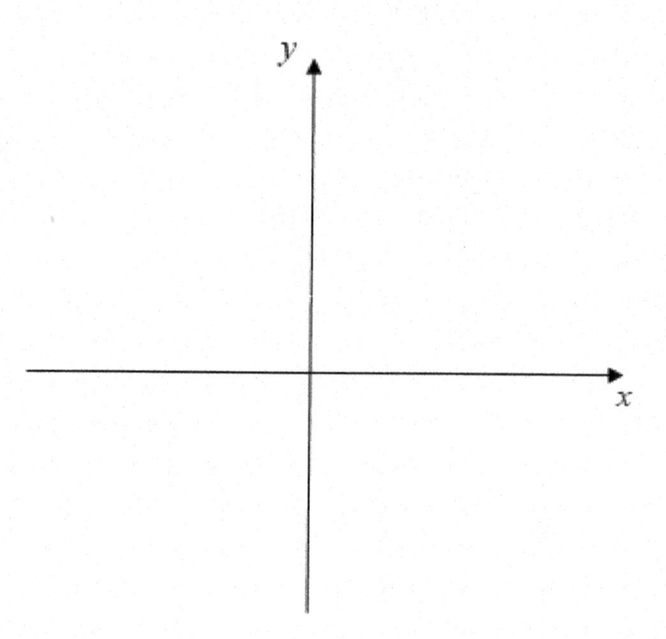

b. $y = \sin(x + \frac{\pi}{2})$

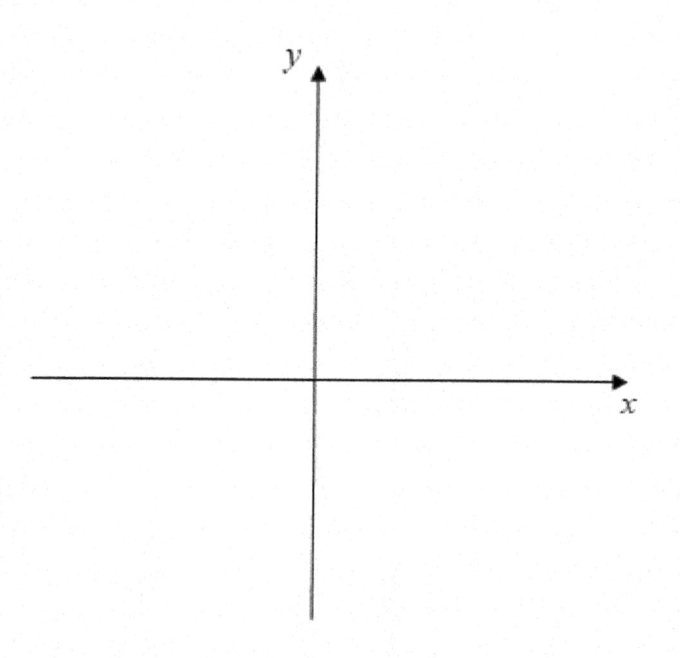

What is the **Relationship between Graphs of Sine and Cosine Functions?**
Fill in the blanks:

$\sin\left(x+\frac{\pi}{2}\right)=$ ______________ and $\cos\left(x-\frac{\pi}{2}\right)=$ ______________

Section 7.2 Objective 2 Sketching Graphs of the Form $y = A\sin(Bx - C)$ and $y = A\cos(Bx - C)$

What are the seven **Steps for Sketching Functions of the Form** $\boldsymbol{y = A\sin(Bx - C)}$ **and** $\boldsymbol{y = A\cos(Bx - C)}$?

Step 1.

Step 2.

Step 3.

Step 4.

Step 5.

Step 6.

Step 7.

Work through the interactive video with Example 2 showing all work below.
Sketch the graph of each function.

a. $y = 3\sin(2x - \pi)$

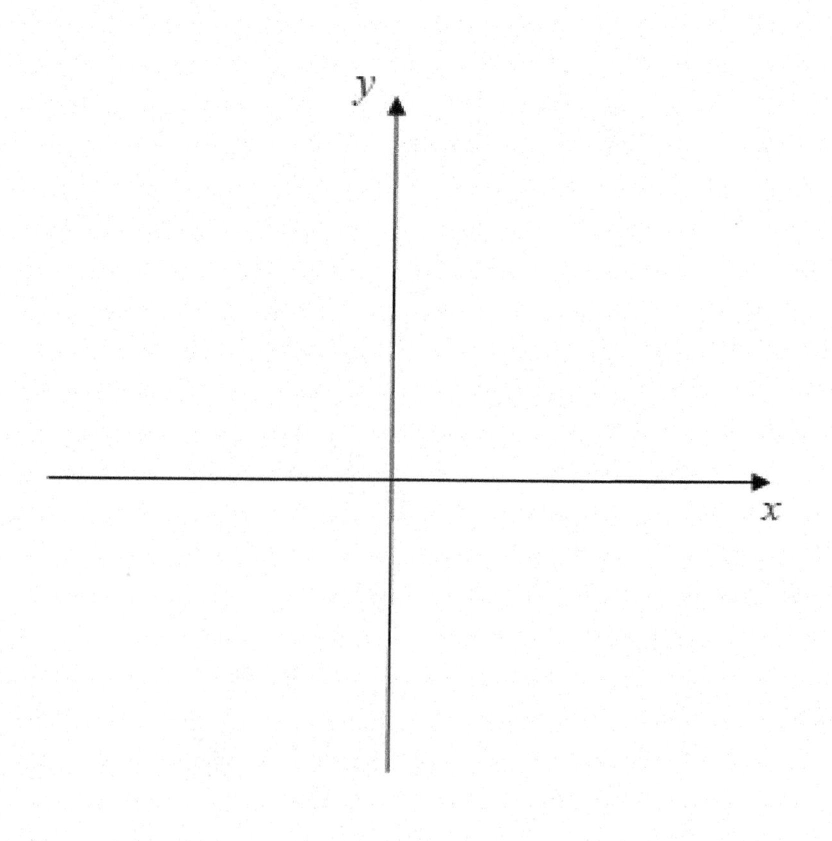

b. $y = -2\cos\left(3x + \frac{\pi}{2}\right)$

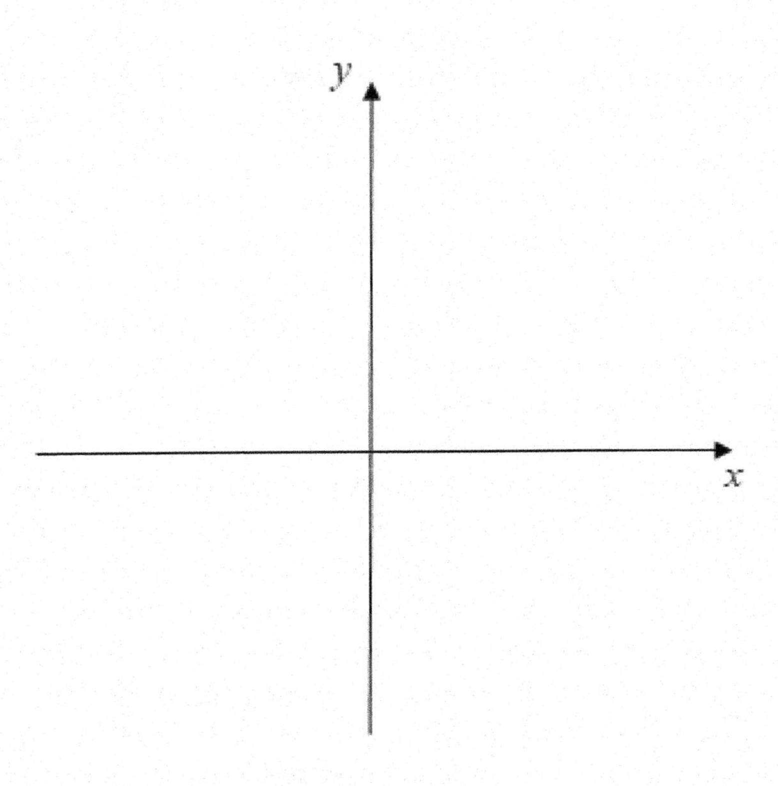

c. $y = 3\sin(\pi - 2x)$

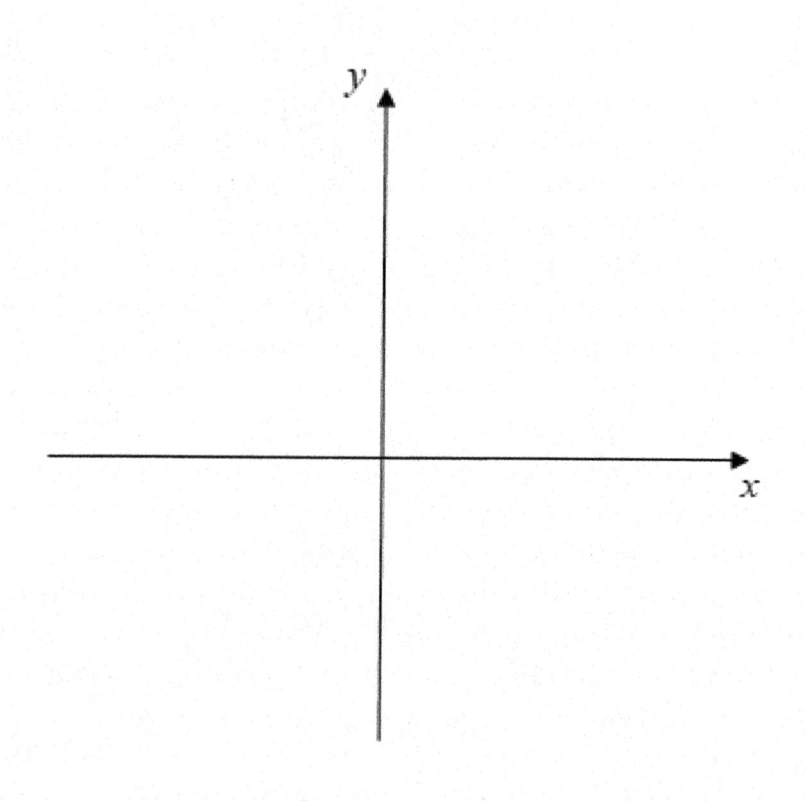

d. $y = -2\cos\left(-3x + \frac{\pi}{2}\right)$

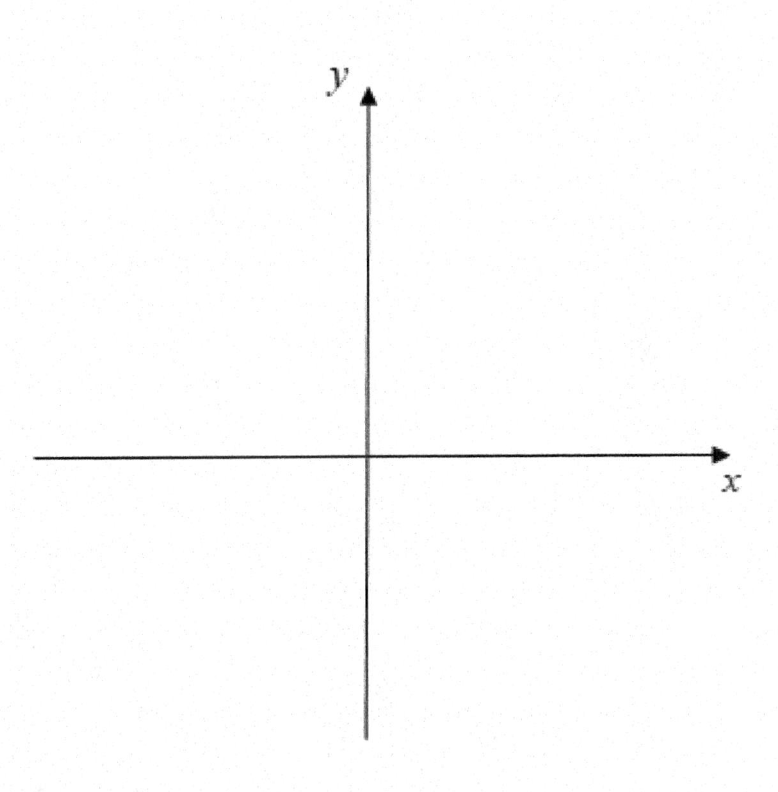

Section 7.2 Objective 3 Sketching Graphs of the Form $y = A\sin(Bx - C) + D$ and $y = A\cos(Bx - C) + D$

What are the seven **Steps for Sketching Functions of the Form** $\boldsymbol{y = A\sin(Bx - C) + D}$ **and** $\boldsymbol{y = A\cos(Bx - C) + D}$?

Step 1.

Step 2.

Step 3.

Step 4.

Step 5.

Step 6.

Step 7.

Work through the interactive video with Example 3 showing all work below.
Sketch the graph of each function.

a. $y = 3\sin\left(2x - \frac{\pi}{2}\right) - 1$

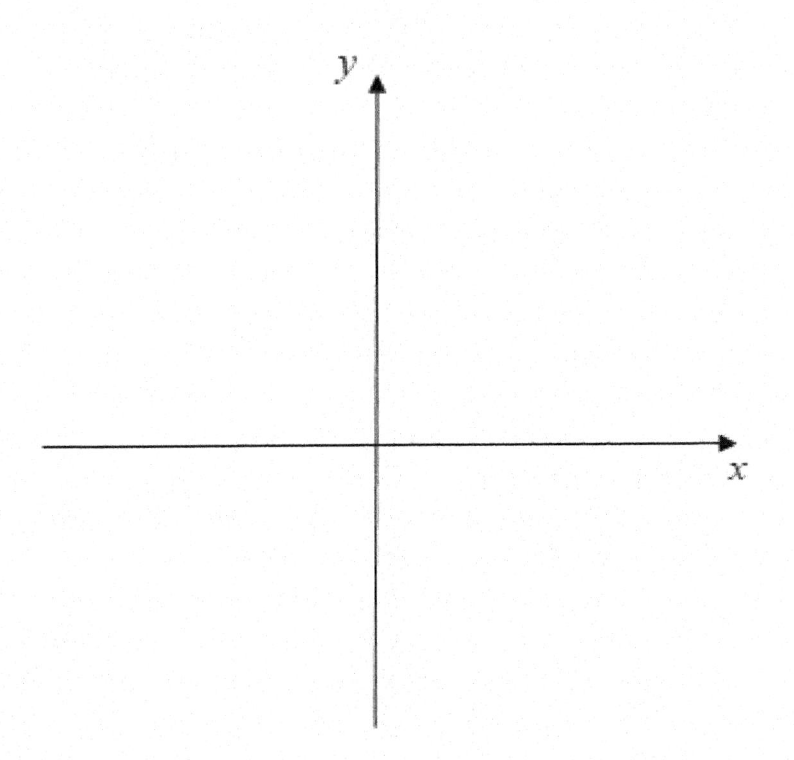

b. $y = 4 - \cos(-\pi x + 2)$

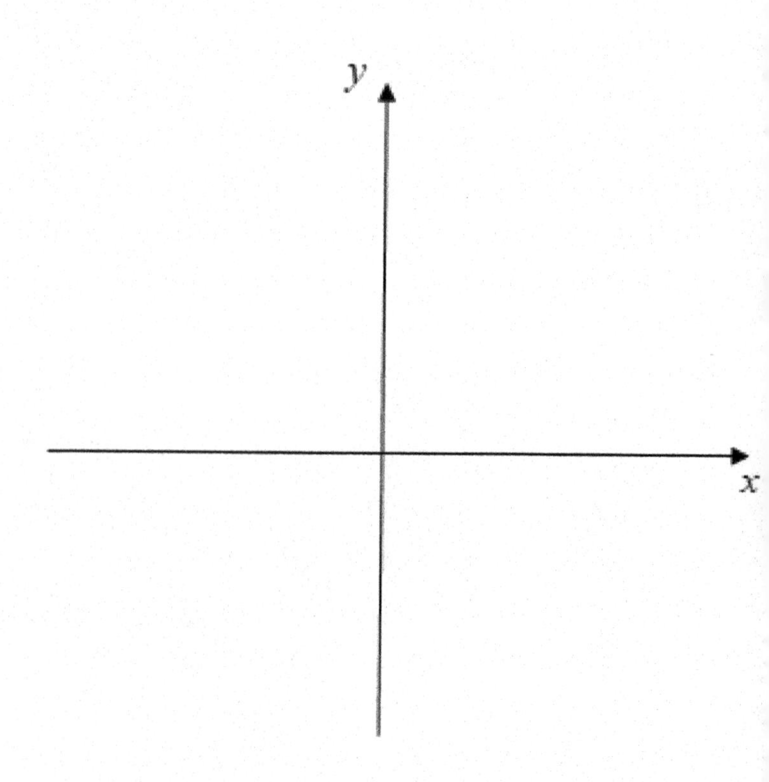

Work through the **Guided Visualization** seen on page 7.2-43. Sketch and label the graphs of $y = A\sin(Bx - C) + D$ and $y = A\cos(Bx - C) + D$. You choose the values of *A, B, C,* and *D.*

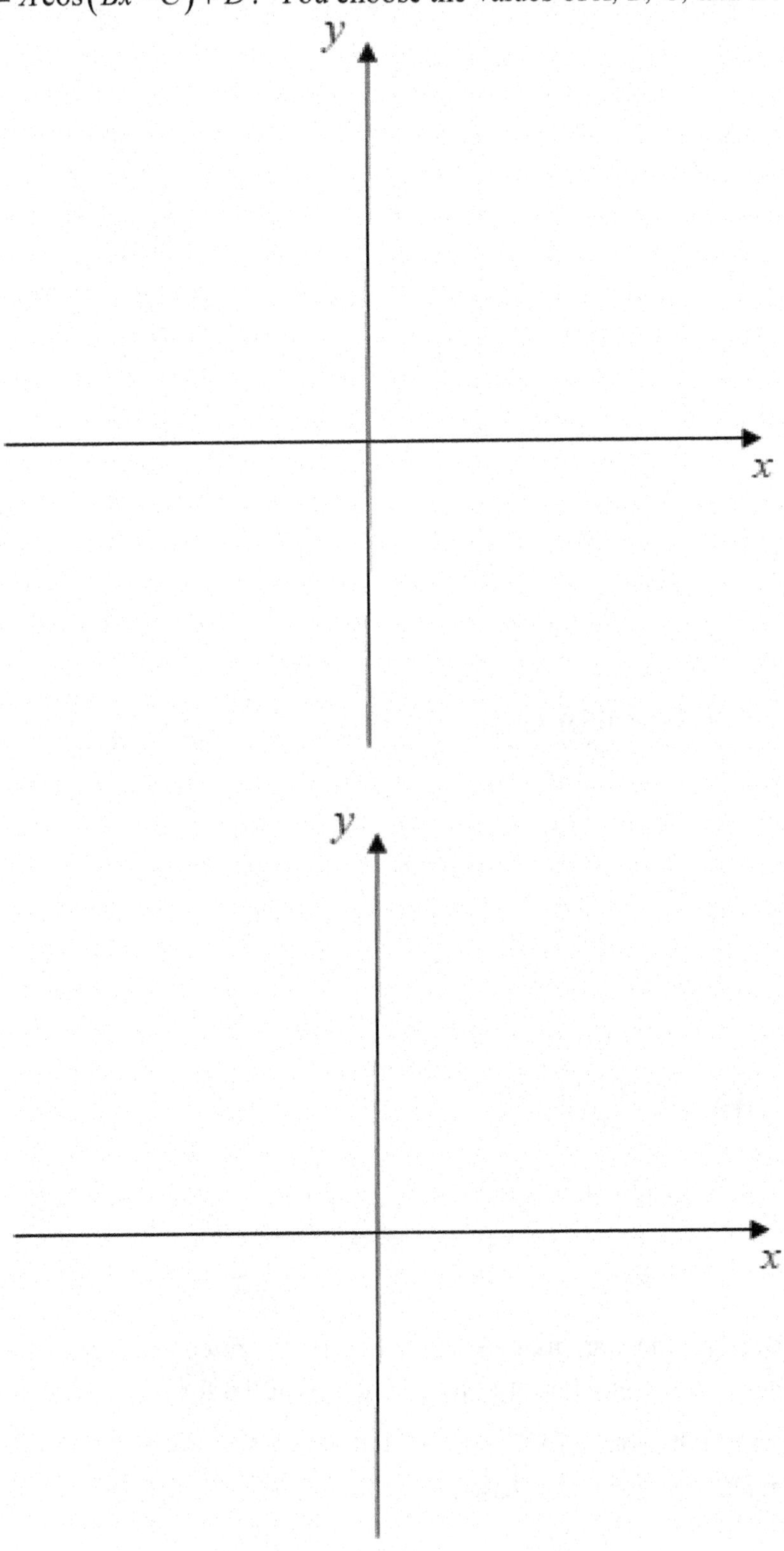

Section 7.2 Objective 4 Determine the Equation of a Function of the Form $y = A\sin(Bx - C) + D$ and $y = A\cos(Bx - C) + D$ Given its Graph

The graph below can be seen in your eText.

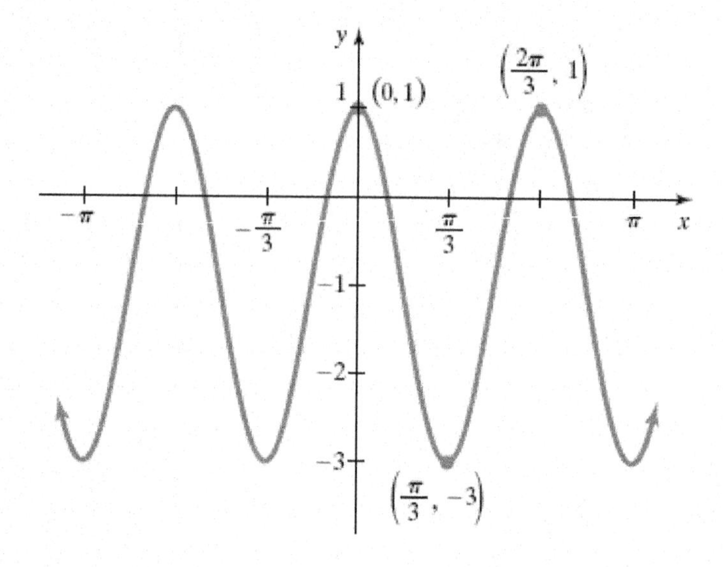

Which 5 functions listed in your eText describe the graph above?

Because there are many ways to describe a given sine or cosine curve, we must be given whether the graph represents a function of the form $y = A\sin(Bx - C) + D$ or of the form $y = A\cos(Bx - C) + D$. We will always also assume that $B > 0$. If these assumptions are made, then we can determine the function by following the six steps that can be seen on the next page.

Write down the six steps necessary for determining an equation of a function of the form $y = A\sin(Bx - C) + D$ or $y = A\cos(Bx - C) + D$ given the graph.

Step 1.

Step 2.

Step 3.

Step 4.

Step 5.

Step 6.

Work through the interactive video that accompanies Example 4.

Example 4a.

The graph of a function of the form $y = A\cos(Bx - C) + D$, where $B > 0$ is given below. The five quarter points of one cycle of the graph are labeled. These five quarter points on the graph correspond to the five quarter points of the graph of $y = \cos x$ over the interval $[0, 2\pi]$. Determine the specific function that is represented by the given graph based on the association of the labeled quarter points and the quarter points of the graph of $y = \cos x$ over the interval $[0, 2\pi]$.

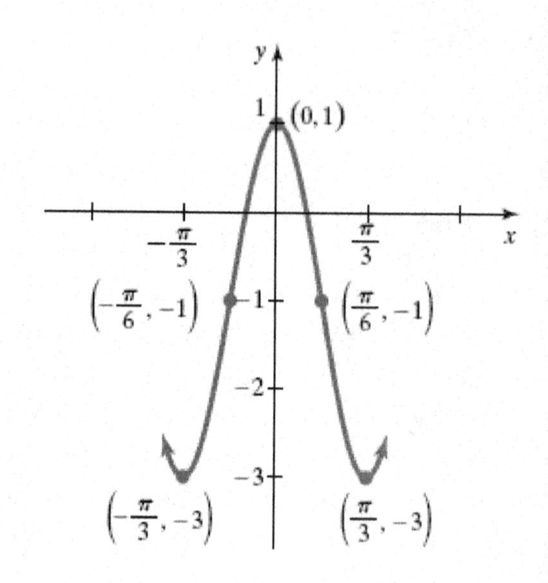

Example 4b.

The graph of a function of the form $y = A\sin(Bx - C) + D$, where $B > 0$ is given below. The five quarter points of one cycle of the graph are labeled. These five quarter points on the graph correspond to the five quarter points of the graph of $y = \sin x$ over the interval $[0, 2\pi]$. Determine the specific function that is represented by the given graph based on the association of the labeled quarter points and the quarter points of the graph of $y = \sin x$ over the interval $[0, 2\pi]$.

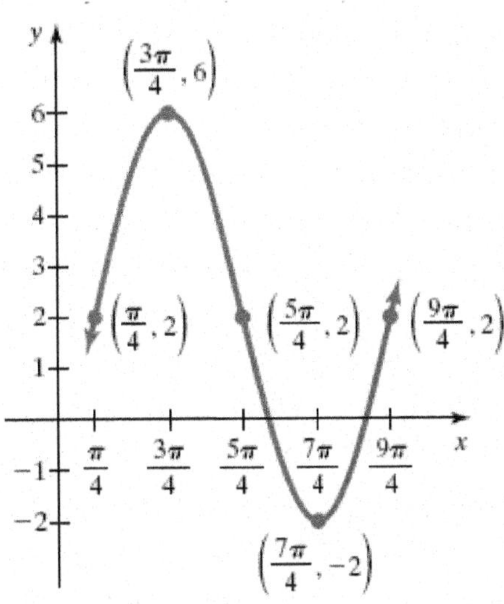

Section 7.3 Guided Notebook

7.3 The Graphs of the Tangent, Cotangent, Secant, and Cosecant Functions

- ☐ Work through Section 7.3 TTK #1–5
- ☐ Work through Section 7.3 Objective 1
- ☐ Work through Section 7.3 Objective 2
- ☐ Work through Section 7.3 Objective 3
- ☐ Work through Section 7.3 Objective 4
- ☐ Work through Section 7.3 Objective 5
- ☐ Work through Section 7.3 Objective 6
- ☐ Work through Section 7.3 Objective 7

Section 7.3 Objective 1 Understanding the Graph of the Tangent Function and Its Properties

Watch the video that accompanies this objective to see how to sketch the graph of $y = \tan x$.

Sketch the graph of the **principal cycle** of $y = \tan x$.

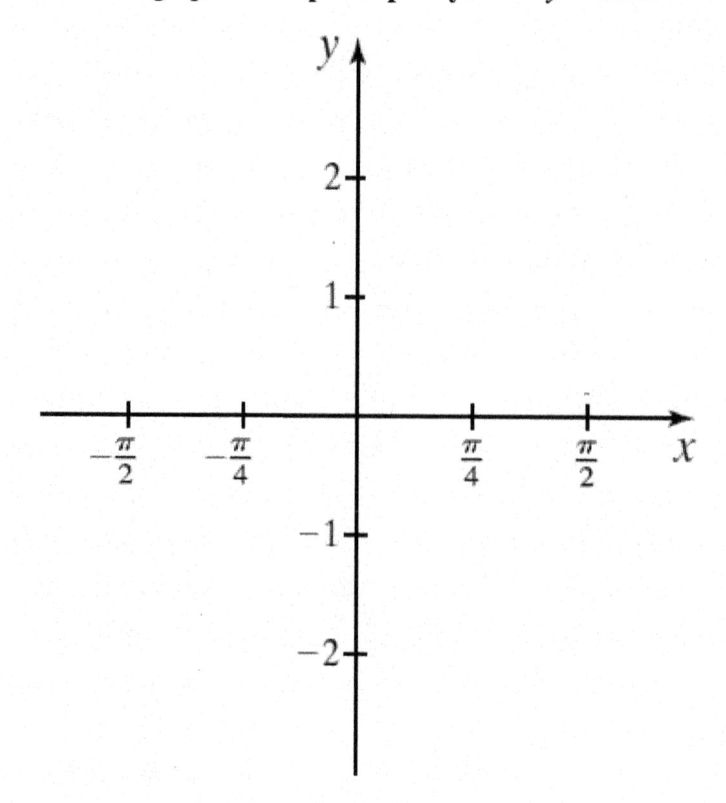

What are the three special points in each cycle of the graph of $y = \tan x$ that will help us to sketch the graph? Go back to the sketch above and add these special points. Also, make sure to draw and label the vertical asymptotes.

Fill in the blanks below:

The principal cycle of the tangent function is defined on the interval: _____

The coordinates of the center point of the principal cycle of the tangent function are _______.

The coordinates of the two halfway points are: _________ and _________.

Write down the **Characteristics of the Tangent Function.**

Work through the video with Example 1 showing all work below.

List all the halfway points of $y = \tan x$ on the interval $\left[-\pi, \frac{5\pi}{2}\right]$ that have a *y*-coordinate of -1.

Section 7.3 Objective 2 Sketching Functions of the Form $y = A\tan(Bx - C) + D$

What are the six **Steps for Sketching Functions of the Form** $y = A\tan(Bx - C) + D$**?**

Step 1.

Step 2.

Step 3.

Step 4.

Step 5.

Step 6.

Work through the interactive video with Example 2 showing all work below.
For each function, determine the interval for the principal cycle. Then for the principal cycle, determine the equations of the vertical asymptotes, the coordinates of the center points, and the coordinates of the halfway points. Sketch the graph.

a. $y = \tan\left(x - \frac{\pi}{6}\right)$

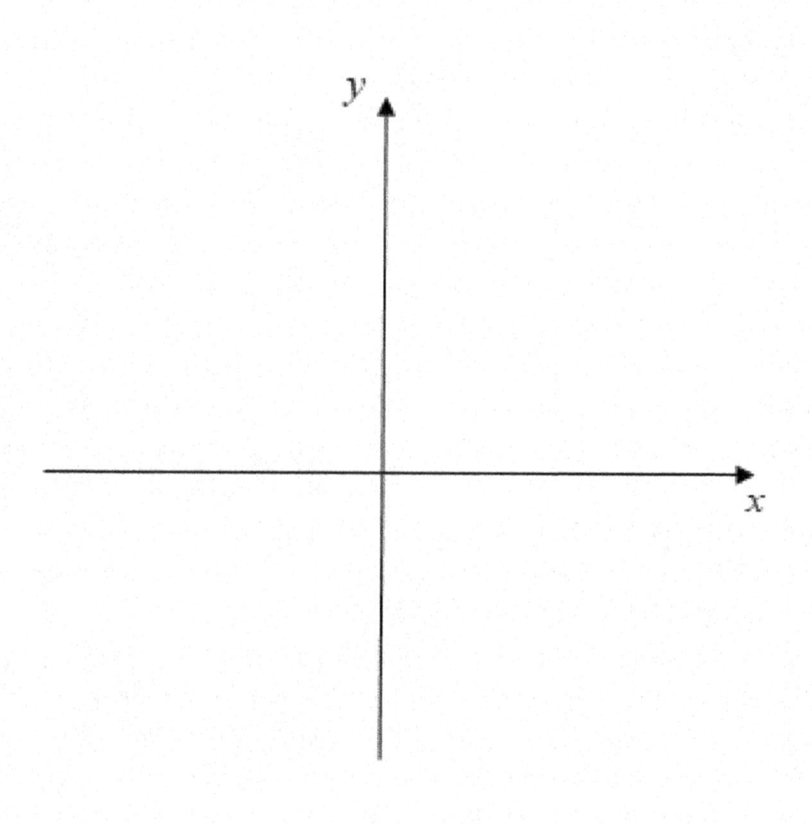

b. $y = 4\tan(\pi - 2x) + 3$

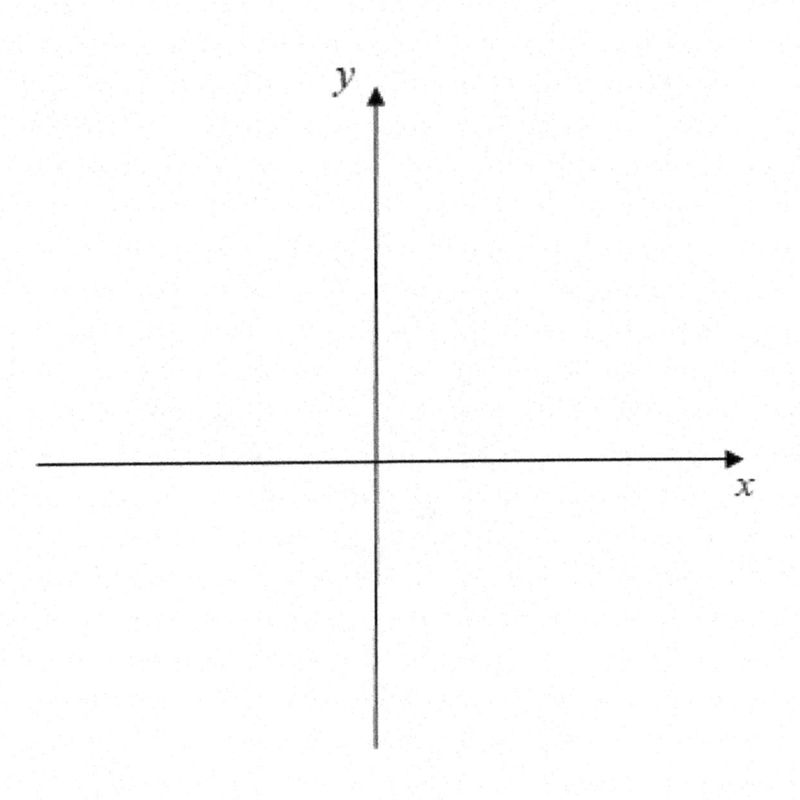

c. $y = \frac{1}{2}\tan(3x) - 1$

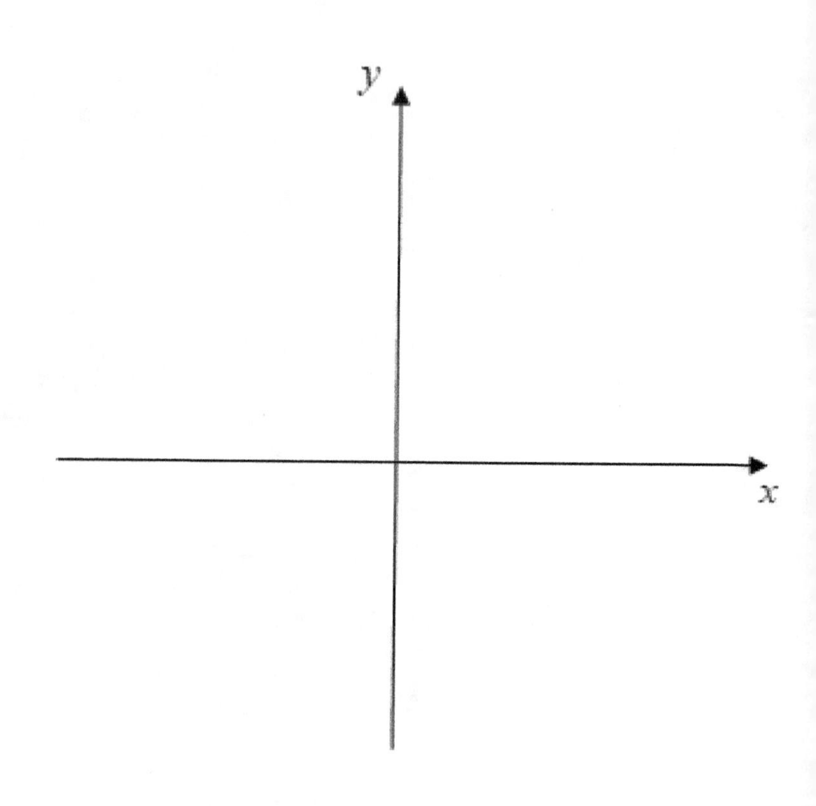

Section 7.3 Objective 3 Understanding the Graph of the Cotangent Function and Its Properties

Watch the video that accompanies this objective to see how to sketch the graph of $y = \cot x$.

Sketch the graph of the **principal cycle** of $y = \cot x$.

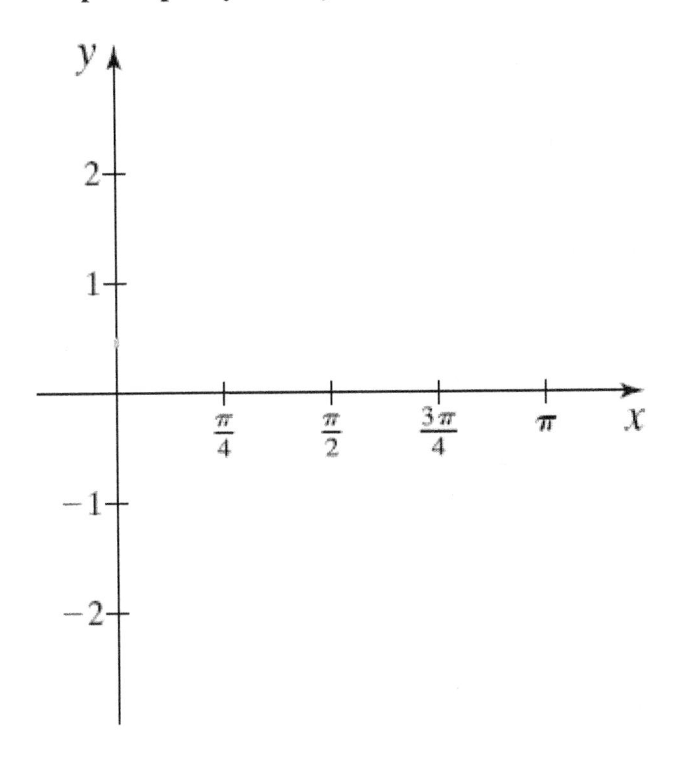

What are the three special points in each cycle of the graph of $y = \cot x$ that will help us to sketch the graph? Go back to the sketch above and add these special points. Also, make sure to draw and label the vertical asymptotes.

Fill in the blanks below:

The principal cycle of the cotangent function is defined on the interval: ________.

The coordinates of the center point of the principal cycle of the cotangent function are ________.

The coordinates of the two halfway points are: ________ and ________.

What are the **Characteristics of the Cotangent Function**?

Work through the video with Example 3 showing all work below.
List all points on the graph of $y = \cot x$ on the interval $[-2\pi, 2\pi]$ that have a y-coordinate of $-\sqrt{3}$.

Section 7.3 Objective 4 Sketching Functions of the Form $y = A\cot(Bx - C) + D$

What are the six **Steps for Sketching Functions of the Form $y = A\cot(Bx - C) + D$?**

Step 1.

Step 2.

Step 3.

Step 4.

Step 5.

Step 6.

Work through the interactive video with Example 4 and show all work below.

For each function, determine the interval for the principal cycle. Then for the principal cycle, determine the equations of the vertical asymptotes, the coordinates of the center points, and the coordinates of the halfway points. Sketch the graph.

a. $y = \cot(2x + \pi) + 1$

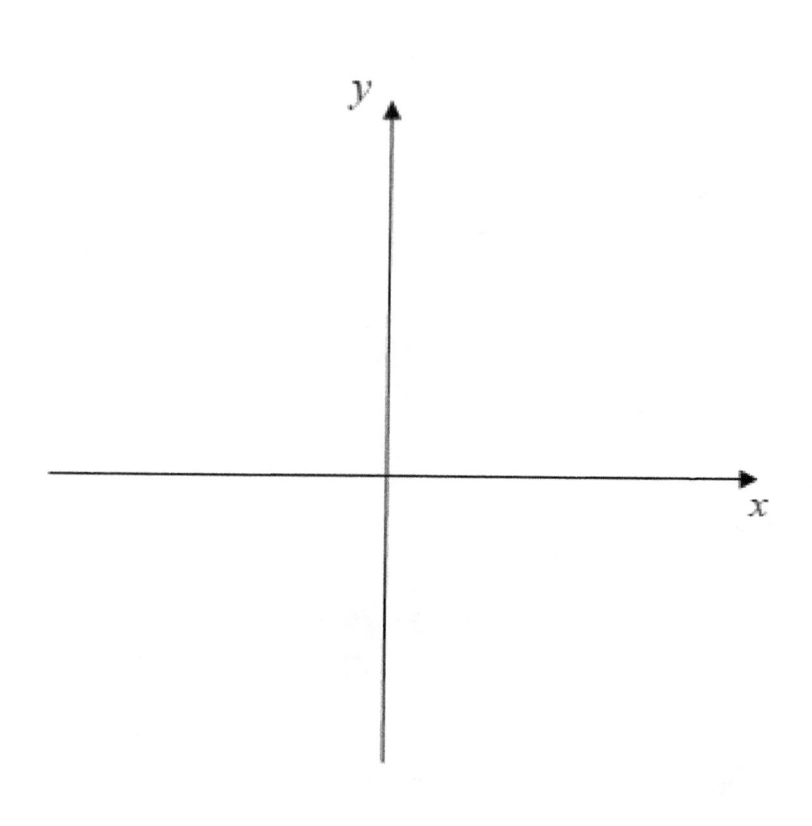

b. $y = -3\cot\left(x - \dfrac{\pi}{4}\right)$

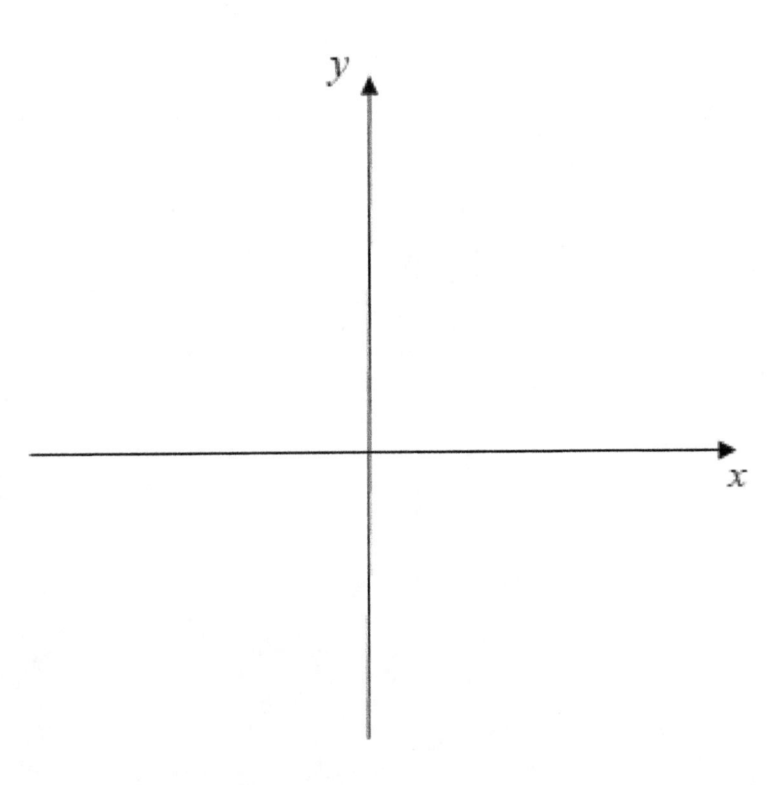

Section 7.3 Objective 5 Determine the Equation of a Function of the Form $y = A\tan(Bx - C) + D$ or $y = A\cot(Bx - C) + D$ Given its Graph

Given a graph of the form $y = A\tan\left(Bx - C\right) + D$ or $y = A\cot\left(Bx - C\right) + D$ for $B > 0$, what five characteristics of the principal cycle of the given graph must we first identify in order to determine the proper function?

1.

2.

3.

4.

5.

Write down the five steps for determining the equation of a function of the form $y = A\tan(Bx - C) + D$ or $y = A\cot(Bx - C) + D$ given the graph.

Step 1.

Step 2.

Step 3.

Step 4.

Step 5.

Work through the interactive video that accompanies Example 5.
The principal cycle of the graphs of two trigonometric functions of the form $y = A\tan(Bx - C) + D$ or $y = A\cot(Bx - C) + D$ for $B > 0$ are given below. Determine the equation of the function represented by each graph.

a. $y = A\tan(Bx - C) + D$

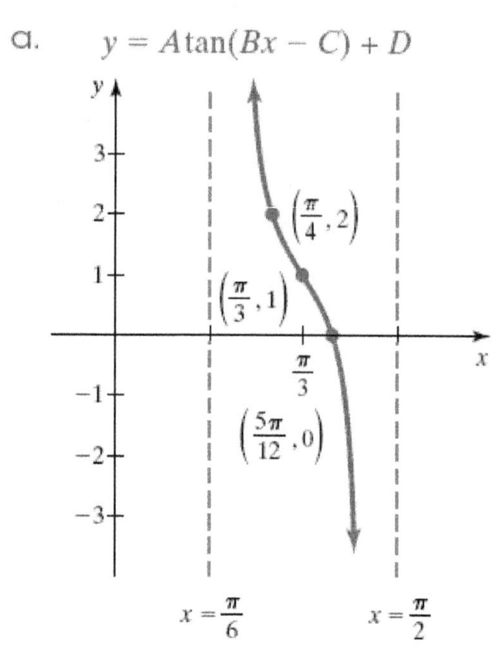

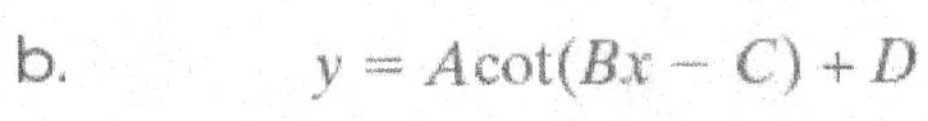

b. $y = A\cot(Bx - C) + D$

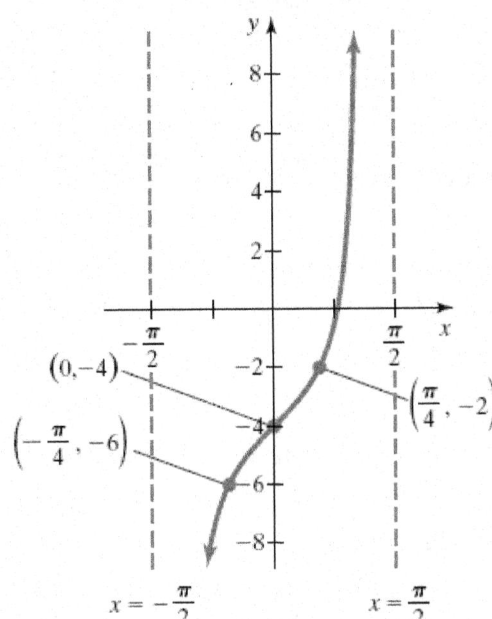

y
8
6
4
2
−2
−4
−6
−8
x
$-\frac{\pi}{2}$
$\frac{\pi}{2}$
$(0, -4)$
$\left(-\frac{\pi}{4}, -6\right)$
$\left(\frac{\pi}{4}, -2\right)$
$x = -\frac{\pi}{2}$
$x = \frac{\pi}{2}$

Section 7.3 Objective 6 Understanding the Graphs of the Cosecant and Secant Functions and Their Properties

Work through the interactive video that accompanies this objective. Click on "The graph of $y = \csc x$" then click on "The graph of $y = \sec x$" and graph each function on the grids below.

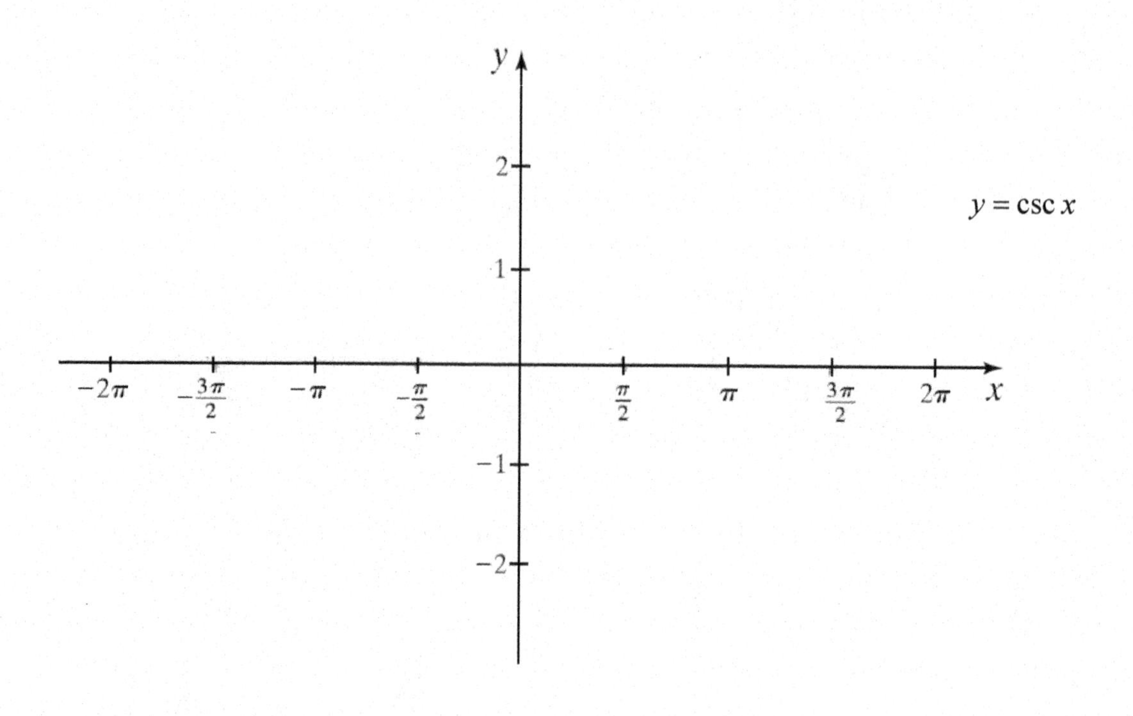

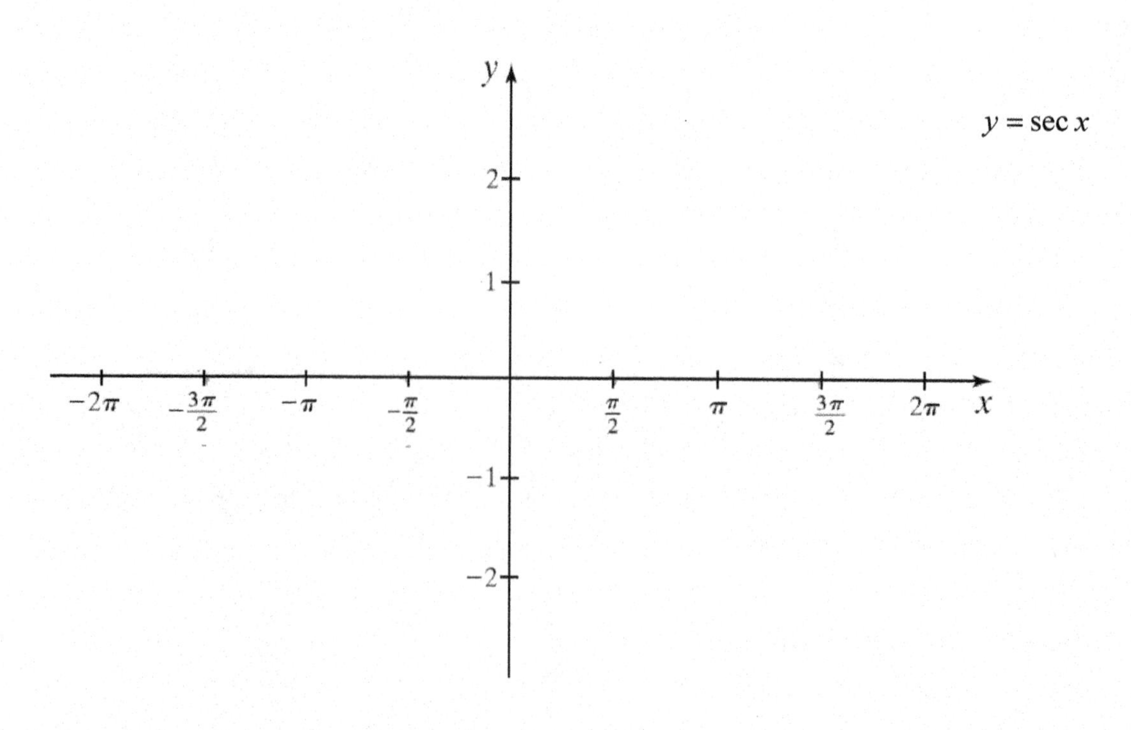

What are the **Characteristics of the Cosecant Function**?

What are the **Characteristics of the Secant Function**?

Section 7.3 Objective 7 Sketching Functions of the Form $y = A\csc(Bx - C) + D$ and $y = A\sec(Bx - C) + D$

What are the four **Steps for Sketching Functions of the Form $y = A\csc(Bx - C) + D$ and $y = A\sec(Bx - C) + D$?**

Step 1.

Step 2.

Step 3.

Step 4.

Work through the interactive video with Example 6 and show all work below.
Determine the equations of the vertical asymptotes and all relative maximum and relative minimum points of two cycles of each function and then sketch its graph.

a. $y = -2\csc(x + \pi)$

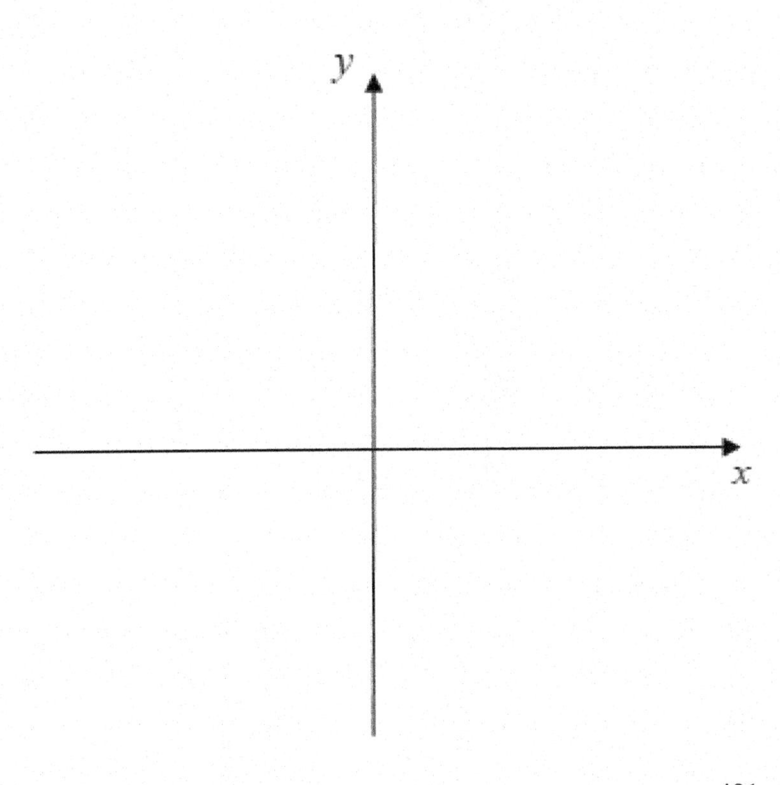

b. $y = -\csc(\pi x) - 2$

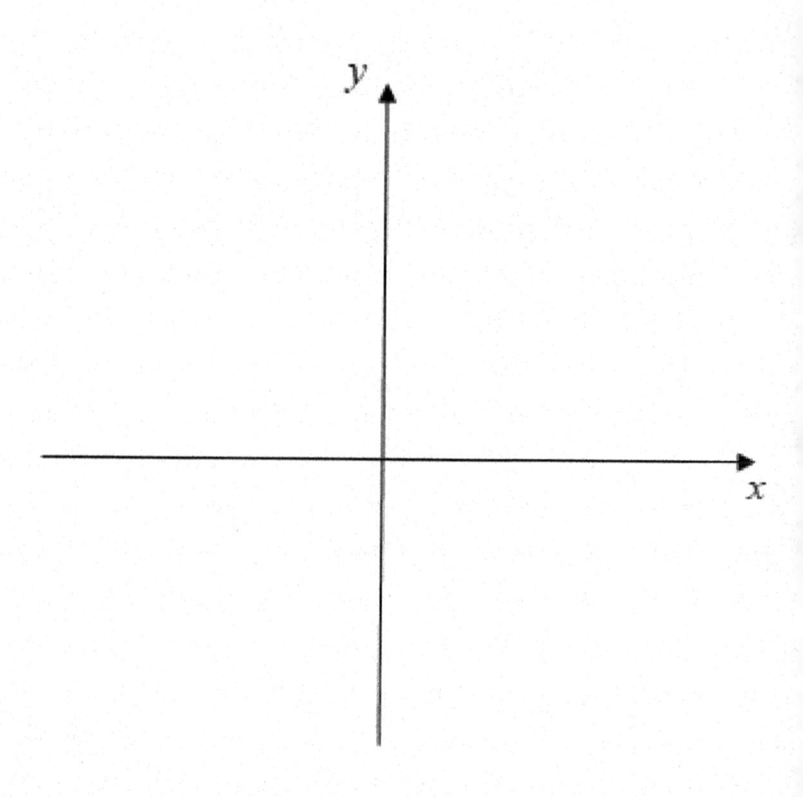

c. $y = 3\sec(\pi - x)$

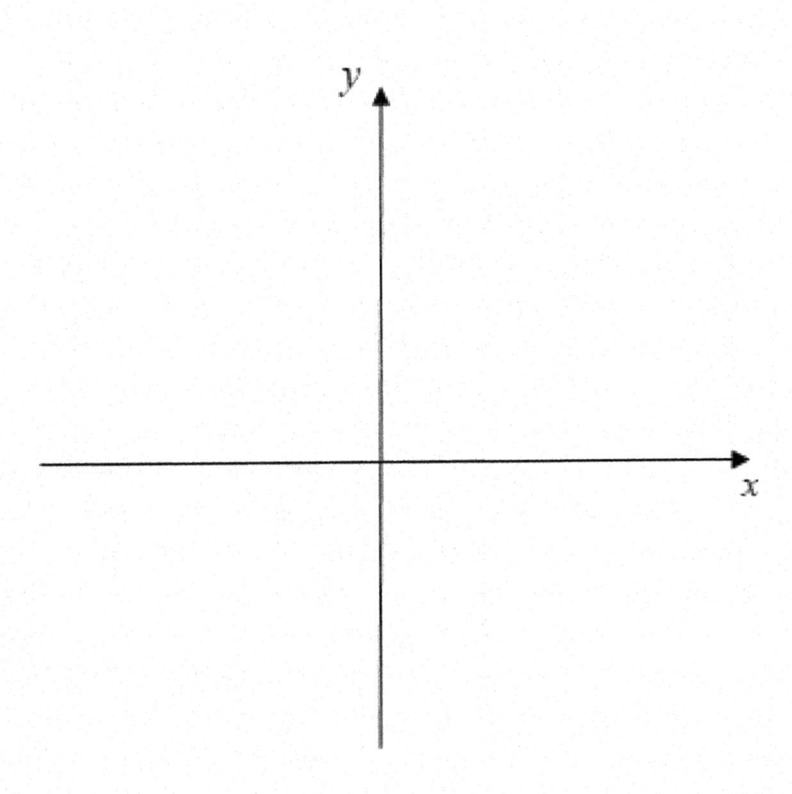

d. $y = \sec\left(2x + \frac{\pi}{2}\right) + 1$

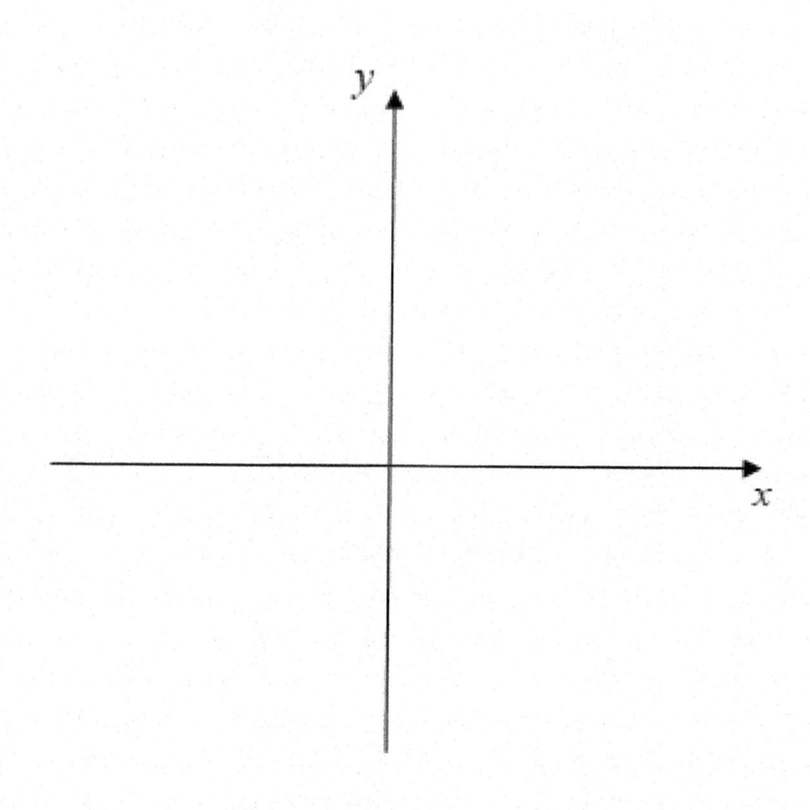

Section 7.4 Guided Notebook

7.4 Inverse Trigonometric Functions I

- [ ] Work through Section 7.4 TTK #1–11
- [ ] Work through Section 7.4 Introduction
- [ ] Work through Section 7.4 Objective 1
- [ ] Work through Section 7.4 Objective 2
- [ ] Work through Section 7.4 Objective 3

Section 7.4 Inverse Trigonometric Functions I

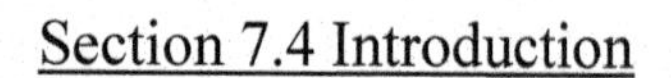

Section 7.4 Introduction

Explain why the function $f(x) = x^2 + 1$ is **not** one-to-one.

How can you restrict the domain of $f(x) = x^2 + 1$ so that it is a one-to-one function?

Watch the animation that accompanies the Introduction and sketch the graph in the animation that is a one-to-one function.

Section 7.4 Objective 1 Understanding and Finding the Exact and Approximate Values of the Inverse Sine Function

Read through this objective and watch the animation that describes the inverse sine function.
Take your animation notes here:

What is the definition of the **Inverse Sine Function**?
(Define in words and sketch the graph.)

What are the four **Steps for Determining the Exact Value of** $\sin^{-1} x$ **?**

Step 1.

Step 2.

Step 3.

Step 4.

Work through the interactive video with Example 1 showing all work below.
Determine the exact value of each expression.

a. $\sin^{-1}\left(\frac{1}{2}\right)$

b. $\sin^{-1}\left(-\frac{\sqrt{3}}{2}\right)$

Work through Example 2 showing all work below.
Use a calculator to approximate each value, or state that the value does not exist.

a. $\sin^{-1}(.7)$

b. $\sin^{-1}(-.95)$

c. $\sin^{-1}(3)$

Section 7.4 Objective 2 Understanding and Finding the Exact and Approximate Values of the Inverse Cosine Function

What is the definition of the **Inverse Cosine Function**? (Define and sketch.)

What are the four **Steps for Determining the Exact Value of** $\cos^{-1} x$ **?**

Step 1.

Step 2.

Step 3.

Step 4.

Work through the interactive video with Example 3 showing all work below.
Determine the exact value of each expression.

a. $\cos^{-1}(1)$

b. $\cos^{-1}\left(-\frac{1}{\sqrt{2}}\right)$

Work through Example 4 showing all work below.
Use a calculator to approximate each value, or state that the value does not exist.

a. $\cos^{-1}(1.5)$

b. $\cos^{-1}(-.25)$

Section 7.4 Objective 3 Understanding and Finding the Exact and Approximate Values of the Inverse Tangent Function

What is the definition the **Inverse Tangent Function**? (Define and sketch.)

What are the four **Steps for Determining the Exact Value of** $\tan^{-1} x$?

Step 1.

Step 2.

Step 3.

Step 4.

Work through the interactive video with Example 5 showing all work below.
Determine the exact value of each expression.

a. $\tan^{-1}\left(\frac{1}{\sqrt{3}}\right)$

b. $\tan^{-1}\left(-\frac{1}{\sqrt{3}}\right)$

Work through Example 6 showing all work below.
Use a calculator to approximate the value of $\tan^{-1}(20)$, or state that the value does not exist.

Section 7.5 Guided Notebook

7.5 Inverse Trigonometric Functions II

- [ ] Work through Section 7.5 TTK #1–9
- [ ] Work through Section 7.5 Objective 1
- [ ] Work through Section 7.5 Objective 2
- [ ] Work through Section 7.5 Objective 3
- [ ] Work through Section 7.5 Objective 4

Section 7.5 Inverse Trigonometric Functions II

Introduction to Section 7.5

Review the graphs of $y = \sin^{-1} x$, $y = \cos^{-1} x$, and $y = \tan^{-1} x$. Sketch these graphs below and state the domain and range of each.

$y = \sin^{-1} x$

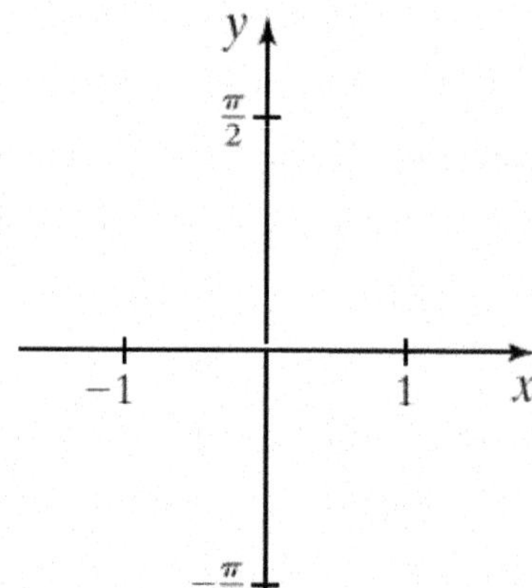

Domain:

Range:

$y = \cos^{-1} x$

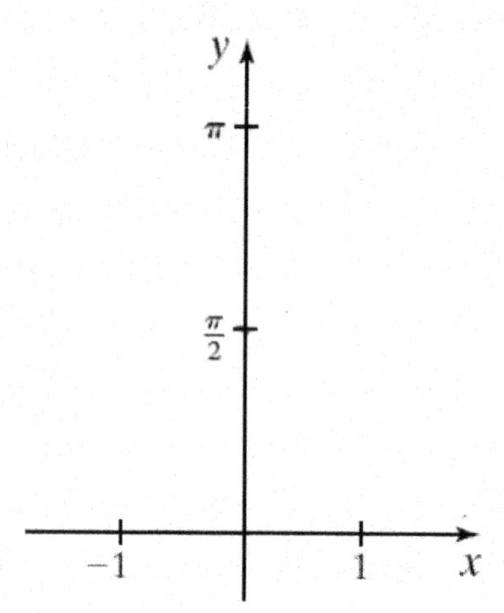

Domain:

Range:

$y = \tan^{-1} x$

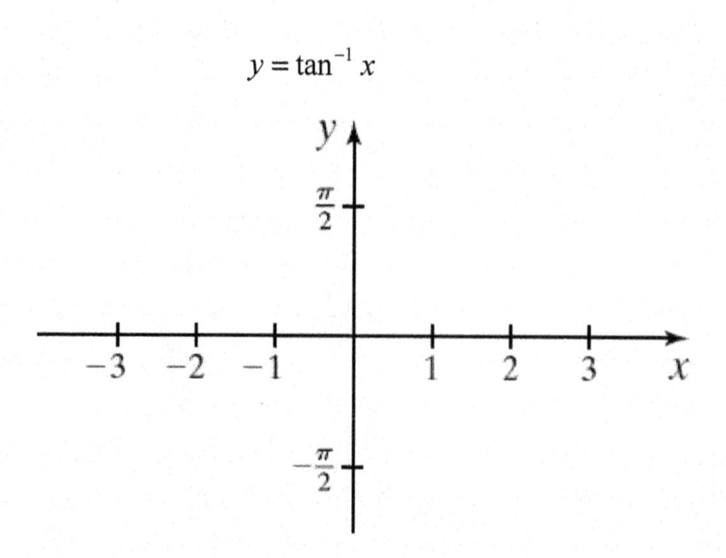

Domain:

Range:

Section 7.5 Objective 1 Evaluating Composite Functions Involving Inverse Trigonometric Functions of the Form $f \circ f^{-1}$ and $f^{-1} \circ f$

What are the **Cancellation Equations for Compositions of Inverse Trigonometric Functions**?

Work through the interactive video with Example 1 showing all work below.
Find the exact value of each expression or state that it does not exist.

a. $\sin\left(\sin^{-1}\frac{1}{2}\right)$

b. $\cos\left(\cos^{-1}\frac{3}{2}\right)$

c. $\tan(\tan^{-1}(8.2))$

d. $\sin(\sin^{-1}(1.3))$

Work through the interactive video with Example 2 showing all work below.
Find the exact value of each expression or state that it does not exist.

a. $\sin^{-1}\left(\sin\frac{\pi}{6}\right)$

b. $\cos^{-1}\left(\cos\frac{2\pi}{3}\right)$

c. $\sin^{-1}\left(\sin\frac{4\pi}{3}\right)$

d. $\tan^{-1}\left(\tan\frac{7\pi}{10}\right)$

Section 7.5 Objective 2 Evaluating Composite Functions Involving Inverse Trigonometric Functions of the Form $f \circ g^{-1}$ and $f^{-1} \circ g$

Work through the interactive video with Example 3 showing all work below.
Find the exact value of each expression or state that it does not exist.

a. $\cos(\tan^{-1}\sqrt{3})$

b. $\csc\left(\cos^{-1}\left(-\frac{\sqrt{3}}{2}\right)\right)$

c. $\sec\left(\sin^{-1}\left(-\frac{\sqrt{5}}{8}\right)\right)$

Work through the interactive video with Example 4 showing all work below.
Find the exact value of each expression or state that it does not exist.

a. $\sin^{-1}\left(\cos\left(-\frac{2\pi}{3}\right)\right)$

b. $\cos^{-1}\left(\sin\frac{\pi}{7}\right)$

Section 7.5 Objective 3 Understanding the Inverse Cosecant, Inverse Secant, and Inverse Cotangent Functions

What is the definition of the **Inverse Cosecant Function**? (Define and sketch.)

What is the definition of the **Inverse Secant Function**? (Define and sketch.)

What is the definition of the **Inverse Cotangent Function**? (Define and sketch.)

Work through Example 5 showing all work below.
Find the exact value of $\sec^{-1}(-\sqrt{2})$ or state that it does not exist.

Work through Example 6 showing all work below.
Use a calculator to approximate each value or state that the value does not exist.

a. $\sec^{-1}(5)$

b. $\cot^{-1}(-10)$

c. $\csc^{-1}(0.4)$

Section 8.1 Guided Notebook

8.1 Trigonometric Identities

- ☐ Work through Section 8.1 TTK #1–3
- ☐ Work through Section 8.1 Objective 1
- ☐ Work through Section 8.1 Objective 2
- ☐ Work through Section 8.1 Objective 3
- ☐ Work through Section 8.1 Objective 4
- ☐ Work through Section 8.1 Objective 5
- ☐ Work through Section 8.1 Objective 6
- ☐ Work through Section 8.1 Objective 7
- ☐ Work through Section 8.1 Objective 8

Section 8.1 Trigonometric Identities

Section 8.1 Objective 1 Reviewing the Fundamental Identities

What are the two **Quotient Identities**?

What are the six **Reciprocal Identities**?

What are the three **Pythagorean Identities**?

What are the four **Odd Properties**?

What are the two **Even Properties**?

Section 8.1 Objective 2 Substituting Known Identities to Verify an Identity

Work through the interactive video with Example 1 showing all work below.
Verify each identity.

a. $\tan x \cot x = 1$

b. $\sec^2 3x + \cot^2 3x - \tan^2 3x = \csc^2 3x$

c. $(5\sin y + 2\cos y)^2 + (5\cos y - 2\sin y)^2 = 29$

Section 8.1 Objective 3 Changing to Sines and Cosines to Verify an Identity

Work through the interactive video with Example 2 showing all work below.
Verify each identity.

a. $\sin^2 t = \tan t \cot t - \cos^2 t$

b. $\dfrac{\sec\theta \csc\theta}{\cot\theta} = \sec^2\theta$

c. $\dfrac{\cos(-\theta)}{\sec\theta} + \sin(-\theta)\csc\theta = -\sin^2\theta$

Section 8.1 Objective 4 Factoring to Verify an Identity

Write down the following special factoring formulas:

Difference of Two Squares:

Perfect Square Formulas:

Sum of Two Cubes:

Difference of Two Cubes:

Work through the interactive video with Example 3 and show all work below.
Verify each trigonometric identity.

a. $\sin x - \cos^2 x \sin x = \sin^3 x$

b. $\dfrac{\tan^3 \alpha - 1}{\tan \alpha - 1} = \sec^2 \alpha + \tan \alpha$

c. $\dfrac{8\sin^2 \theta - 2\sin \theta - 3}{1 + 2\sin \theta} = 4\sin \theta - 3$

Section 8.1 Objective 5 Separating a Single Quotient into Multiple Quotients to Verify an Identity

When one side of a trigonometric identity is a quotient of the form $\frac{A+B}{C}$, where C is a single trigonometric expression, then we can separate the expression into two expressions using the property:

$$\frac{A+B}{C} =$$

(Fill in the right side of the equals sign above.)

Work through the video with Example 4 and show all work below.

Verify the trigonometric identity $\frac{\sin\alpha+\cos\alpha}{\cos\alpha}-\frac{\sin\alpha+\cos\alpha}{\sin\alpha}=\tan\alpha-\cot\alpha$.

Section 8.1 Objective 6 Combining Fractional Expressions to Verify an Identity

Work through the video with Example 5 and show all work below.

Verify the trigonometric identity $\frac{1-\csc\theta}{\cot\theta}-\frac{\cot\theta}{1-\csc\theta}=2\tan\theta$.

Section 8.1 Objective 7 Multiplying by Conjugates to Verify Identities

Given the expression $A+B$, then the conjugate is ____________________.

Work through the video with Example 6 and show all work below.

Verify the trigonometric identity $\dfrac{\sin\theta}{\csc\theta+1}=\dfrac{1-\sin\theta}{\cot^2\theta}$.

Section 8.1 Objective 8 Summarizing the Techniques for Verifying Identities

What are the six techniques described in **A Summary for Verifying Trigonometric Identities**?

1.

2.

3.

4.

5.

6.

Work through the interactive video with Example 7 and show all work below.
Verify each trigonometric identity.

a. $$\frac{\sin^2 t + 6\sin t + 9}{\sin t + 3} = \frac{3\csc t + 1}{\csc t}$$

b. $\dfrac{2\csc\theta}{\sec\theta}+\dfrac{\cos\theta}{\sin\theta}=3\cot\theta$

c. $\dfrac{1-\sin\theta}{\cos\theta}+\dfrac{\cos\theta}{1-\sin\theta}=2\sec\theta$

Section 8.2 Guided Notebook

8.2 The Sum and Difference Formulas

- ☐ Work through Section 8.2 TTK #1–4
- ☐ Work through Section 8.2 Objective 1
- ☐ Work through Section 8.2 Objective 2
- ☐ Work through Section 8.2 Objective 3
- ☐ Work through Section 8.2 Objective 4
- ☐ Work through Section 8.2 Objective 5

Section 8.2 The Sum and Difference Formulas

Section 8.2 Objective 1 Understanding the Sum and Difference Formulas for the Cosine Function

What are the **Sum and Difference Formulas for the Cosine Function**?

Work through the interactive video with Example 1 showing all work below.
Find the exact value of each trigonometric expression without the use of a calculator.

a. $\cos\left(\frac{2\pi}{3}+\frac{3\pi}{4}\right)$

b. $\cos(225° - 150°)$

Work through the interactive video with Example 2 showing all work below.
Find the exact value of each trigonometric expression without the use of a calculator.

a. $\cos\left(\frac{7\pi}{12}\right)\cos\left(\frac{5\pi}{12}\right)+\sin\left(\frac{7\pi}{12}\right)\sin\left(\frac{5\pi}{12}\right)$

b. $\cos\left(\frac{7\pi}{12}\right)$

c. $\cos(-75°)$

Work through the video with Example 3 showing all work below.
Suppose that the terminal side of angle α lies in Quadrant IV and the terminal side of angle β lies in Quadrant III. If $\cos\alpha = \frac{4}{7}$ and $\sin\beta = -\frac{8}{13}$, find the exact value of $\cos(\alpha+\beta)$.

Section 8.2 Objective 2 Understanding the Sum and Difference Formulas for the Sine Function

What are the two **Cofunction Identities for Sine and Cosine**?

What are the **Sum and Difference Formulas for the Sine Function**?

Work through the interactive video with Example 4 showing all work below.
Find the exact value of each trigonometric expression without the use of a calculator.

a. $\sin\left(-\frac{\pi}{3}+\frac{5\pi}{4}\right)$

b. $\sin(12°)\cos(78°)+\cos(12°)\sin(78°)$

c. $\sin(15°)$

Work through the video with Example 5 showing all work below.

Suppose that α is an angle such that $\tan\alpha=\frac{5}{7}$ and $\cos\alpha<0$. Also, suppose that β is an angle such that $\sec\beta=-\frac{4}{3}$ and $\csc\beta>0$. Find the exact value of $\sin(\alpha+\beta)$.

Section 8.2 Objective 3 Understanding the Sum and Difference Formulas for the Tangent Function

What are the two **Sum and Difference Formulas for the Tangent Function**?

Work through the interactive video with Example 6 showing all work below.
Find the exact value of each trigonometric expression without the use of a calculator.

a. $\tan\left(\frac{5\pi}{6}+\frac{3\pi}{4}\right)$

b. $\tan\left(\frac{\pi}{12}\right)$

Section 8.2 Objective 4 Using the Sum and Difference Formulas to Verify Identities

Work through Example 7 and show all work below.
Verify the trigonometric identity $\sin(2\theta) = 2\sin\theta\cos\theta$.

Work through the video with Example 8 and show all work below.
Verify the trigonometric identity $\csc(\alpha - \beta) = \dfrac{\sin\alpha\cos\beta + \cos\alpha\sin\beta}{\sin^2\alpha - \sin^2\beta}$.

Section 8.2 Objective 5 Using the Sum and Difference Formulas to Evaluate Expressions Involving Inverse Trigonometric Functions

Work through the video with Example 9 and show all work below.

Find the exact value of the expression $\cos\left(\sin^{-1}\left(\frac{1}{5}\right)+\cos^{-1}\left(-\frac{3}{4}\right)\right)$ without using a calculator.

Section 8.3 Guided Notebook

8.3 The Double-Angle and Half-Angle Formulas

- ☐ Work through Section 8.3 TTK #1–6
- ☐ Work through Section 8.3 Objective 1
- ☐ Work through Section 8.3 Objective 2
- ☐ Work through Section 8.3 Objective 3
- ☐ Work through Section 8.3 Objective 4
- ☐ Work through Section 8.3 Objective 5

Section 8.3 The Double-Angle and Half-Angle Formulas

Section 8.3 Objective 1 Understanding the Double-Angle Formulas

What are the three **Double-Angle Formulas**?

What are the three **Double-Angle Formulas for Cosine**?

Work through the interactive video with Example 1 showing all work below. Rewrite each expression as the sine, cosine, or tangent of a double angle. Then evaluate the expression without using a calculator.

a. $\cos^2\left(\frac{11\pi}{12}\right)-\sin^2\left(\frac{11\pi}{12}\right)$

b. $2\sin 67.5°\cos 67.5°$

c. $\dfrac{2\tan\left(-\frac{\pi}{8}\right)}{1-\tan^2\left(-\frac{\pi}{8}\right)}$

d. $2\cos^2 105°-1$

Work through the interactive video with Example 2 showing all work below.

Suppose that the terminal side of an angle θ lies in Quadrant II such that $\sin\theta = \frac{5}{7}$.

Find the values of $\sin 2\theta$, $\cos 2\theta$, and $\tan 2\theta$.

Work through the video with Example 3 showing all work below.

If $\tan 2\theta = -\frac{24}{7}$ for $\frac{3\pi}{2} < 2\theta < 2\pi$, then find the values of $\sin\theta$, $\cos\theta$, and $\tan\theta$.

Section 8.3 Objective 2 Understanding the Power Reduction Formulas

What are the three **Power Reduction Formulas**?

Work through the video with Example 4 showing all work below.
Rewrite the function $f(x) = 6\sin^4 x$ as an equivalent function containing only cosine terms raised to a power of 1.

Section 8.3 Objective 3 Understanding the Half-Angle Formulas

What are the **Half-Angle Formulas for Sine and Cosine**?

What are the **Half-Angle Formulas for Tangent**?

Work through Example 5 showing all work below.
Use a half-angle formulas to evaluate each expression without using a calculator.

a. $\sin\left(-\frac{7\pi}{8}\right)$

b. $\cos(-15°)$

c. $\tan\left(\dfrac{11\pi}{12}\right)$

Work through the interactive video with Example 6 showing all work below.

Suppose that $\csc\alpha = \frac{8}{3}$ such that $\frac{\pi}{2} < \alpha < \pi$.

Find the values of $\sin\left(\frac{\alpha}{2}\right)$, $\cos\left(\frac{\alpha}{2}\right)$, and $\tan\left(\frac{\alpha}{2}\right)$.

Section 8.3 Objective 4 Using the Double-Angle, Power Reduction, and Half-Angle Formulas to Verify Identities

Work through Example 7 and show all work below.

Verify the trigonometric identity $\frac{\cos(2\theta)}{1+\sin(2\theta)} = \frac{\cos\theta - \sin\theta}{\cos\theta + \sin\theta}$.

Section 8.3 Objective 5 Using the Double-Angle and Half-Angle Formulas to Evaluate Expressions Involving Inverse Trigonometric Functions

Work through the video with Example 8 and show all work below.

Find the exact value of the expression $\cos\left(\frac{1}{2}\sin^{-1}\left(-\frac{3}{11}\right)\right)$ without the use of a calculator.

Section 8.4 Guided Notebook

8.4 The Product-to-Sum and Sum-to-Product Formulas

- [ ] Work through Section 8.4 TTK #1–3
- [ ] Work through Section 8.4 Objective 1
- [ ] Work through Section 8.4 Objective 2
- [ ] Work through Section 8.4 Objective 3

Section 8.4 The Product-to-Sum and Sum-to-Product Formulas

Section 8.4 Objective 1 Understanding the Product-to-Sum Formulas

What are the four **Product-to-Sum Formulas**?

Work through the interactive video with Example 1 showing all work below.
Write each product as a sum or difference containing only sines or cosines.

a. $\sin 4\theta \sin 2\theta$

b. $\cos\left(\frac{19\theta}{2}\right)\sin\left(\frac{\theta}{2}\right)$

c. $\cos 11\theta \cos 5\theta$

d. $\sin 6\theta \cos 3\theta$

Work through the video with Example 2 showing all work below.

Determine the exact value of the expression $\sin\left(\frac{3\pi}{8}\right)\cos\left(\frac{\pi}{8}\right)$ without the use of a calculator.

Section 8.4 Objective 2 Understanding the Sum-to-Product Formulas

What are the four **Sum-to-Product Formulas**?

Work through the interactive video with Example 3 showing all work below.
Write each sum or difference as a product of sines and/or cosines.

a. $\sin 5\theta + \sin 3\theta$

b. $\cos\left(\frac{3\theta}{2}\right) - \cos\left(\frac{17\theta}{2}\right)$

Work through the interactive video with Example 4 showing all work below.

Determine the exact value of the expression $\sin\left(\frac{\pi}{12}\right)-\sin\left(\frac{17\pi}{12}\right)$ without the use of a calculator.

Section 8.4 Objective 3 Using the Product-to-Sum and Sum-to-Product Formulas to Verify Identities

Work through Example 5 showing all work below.

Verify the trigonometric identity $\frac{\cos\theta + \cos 3\theta}{2\cos 2\theta} = \cos\theta$.

Section 8.5 Guided Notebook

8.5 Trigonometric Equations

- ☐ Work through Section 8.5 TTK #1–4
- ☐ Work through Section 8.5 Objective 1
- ☐ Work through Section 8.5 Objective 2
- ☐ Work through Section 8.5 Objective 3
- ☐ Work through Section 8.5 Objective 4
- ☐ Work through Section 8.5 Objective 5

Section 8.5 Trigonometric Equations

Section 8.5 Introduction

What is the difference between **identities** and **conditional trigonometric equations**?

What is the difference between **general solution(s)** and **specific solution(s)**?

Watch the animation seen in the intro.
First, sketch the functions $y = \sin\theta$ and $y = b$ for $-1 \le b \le 1$ and show at least 10 solutions to the equation $\sin\theta = b$.

Continue to watch the animation seen in the intro.

How many solutions are there to the equation $\sin\theta = b$ for $b > 1$? Draw a picture and explain your answer.

How many solutions are there to the equation $\sin\theta = b$ for $b < -1$? Draw a picture and explain your answer.

How many solutions are there to the equation $\sin\theta = b$ for $-1 \le b \le 1$ on the interval $0 \le \theta < 2\pi$? Draw a picture and explain your answer.

Section 8.5 Objective 1 Solving Trigonometric Equations That Are Linear in Form

What is a trigonometric equation that is linear in form?

Give at least 3 examples of trigonometric equations that are linear in form:

What are the four **Steps for Solving Trigonometric Equations That Are Linear in Form**?

Step 1.

Step 2.

Step 3.

Step 4.

Work through the interactive video with Example 1 showing all work below.
Determine a general formula (or formulas) for all solutions to each equation.
Then, determine the specific solutions (if any) on the interval $[0, 2\pi)$.

a. $\sin\theta = \frac{1}{2}$

b. $\sqrt{3}\tan\theta + 1 = 0$

c. $\sec\theta = -1$

Work through the interactive video with Example 2 showing all work below. Determine a general formula (or formulas) for all solutions to each equation. Then, determine the specific solutions (if any) on the interval $[0, 2\pi)$.

a. $\sqrt{2}\cos 2\theta + 1 = 0$

b. $\sin\frac{\theta}{2}=-\frac{\sqrt{3}}{2}$

c. $\tan\left(\theta+\frac{\pi}{6}\right)+1=0$

Section 8.5 Objective 2 Solving Trigonometric Equations That Are Quadratic in Form

What is a trigonometric equation that is quadratic in form?

Work through the interactive video with Example 3 showing all work below. Determine a general formula (or formulas) for all solutions to each equation. Then, determine the specific solutions (if any) on the interval $[0, 2\pi)$.

a. $\sin^2\theta - 4\sin\theta + 3 = 0$

b. $4\cos^2\theta - 3 = 0$

Section 8.5 Objective 3 Solving Trigonometric Equations Using Identities

Work through the interactive video with Example 4 showing all work below. Determine a general formula (or formulas) for all solutions to each equation. Then, determine the specific solutions (if any) on the interval $[0, 2\pi)$.

a. $2\sin^2\theta = 3\cos\theta + 3$

b. $\sin\theta\cos\theta = -\frac{1}{2}$

c. $\cos 2\theta + 4\sin^2 \theta = 2$

d. $\sin 5\theta + \sin 3\theta = 0$

Section 8.5 Objective 4 Solving Other Types of Trigonometric Equations

Work through the interactive video with Example 5 and show all work below. Determine a general formula (or formulas) for all solutions to each equation. Then, determine the specific solutions (if any) on the interval $[0, 2\pi)$.

a. $\sin 2\theta + 2\cos\theta \sin 2\theta = 0$

b. $\cos^2\theta = \sin\theta\cos\theta$

c. $\sin\theta + \cos\theta = 1$

Before we can attempt to multiply or divide both sides of a trigonometric equation by a function, what must we first verify? (Hint: Look in the Caution text that follows the solution to Example 5b.)

Section 8.5 Objective 5 Solving Trigonometric Equations Using a Calculator

Work through Example 6 and show all work below.
Approximate all solutions to the trigonometric equation $\cos\theta = -0.4$ on the interval $[0, 2\pi)$. Round your answer to four decimal places.

Why can't we simply use an inverse trigonometric function key on our calculator to solve this trigonometric equation? (Hint: See the Caution text that follows Example 6.)

Section 9.1 Guided Notebook

9.1 Right Triangle Applications

- ☐ Work through Section 9.1 TTK #1 and 2
- ☐ Work through Section 9.1 Objective 1
- ☐ Work through Section 9.1 Objective 2

Section 9.1 Right Triangle Applications

Section 9.1 Objective 1 Solving Right Triangles

Fill in the blanks below:

The goal when solving a right triangle is to determine the _______________ of all angles and the length of __________________________ given certain information.

Work through Example 1 showing all work below.
Solve the given right triangle. Round the length of side a and the measure of the two acute angles to two decimal places.

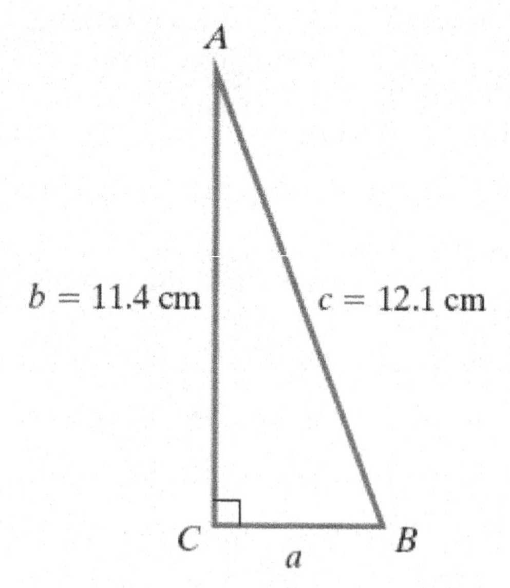

Work through the video with Example 2 showing all work below.
Suppose that the length of the hypotenuse of a right triangle is 11 inches. If one of the acute angles is 37.34°, find the length of the two legs and the measure of the other acute angle. Round all values to two decimal places.

Try solving the right triangles below on your own:

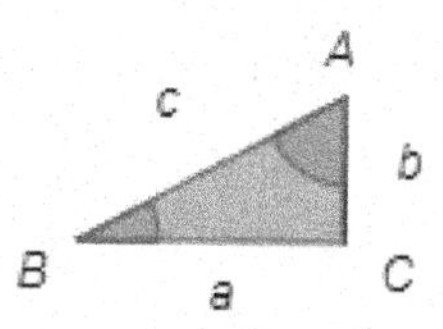

1. $a = 4.9$, $b =$ __________, $c =$ __________
   $A =$ __________, $B = 27^\circ$

2. $a =$ __________, $b =$ __________, $c = 44.6$
   $A = 67^\circ$, $B =$ __________

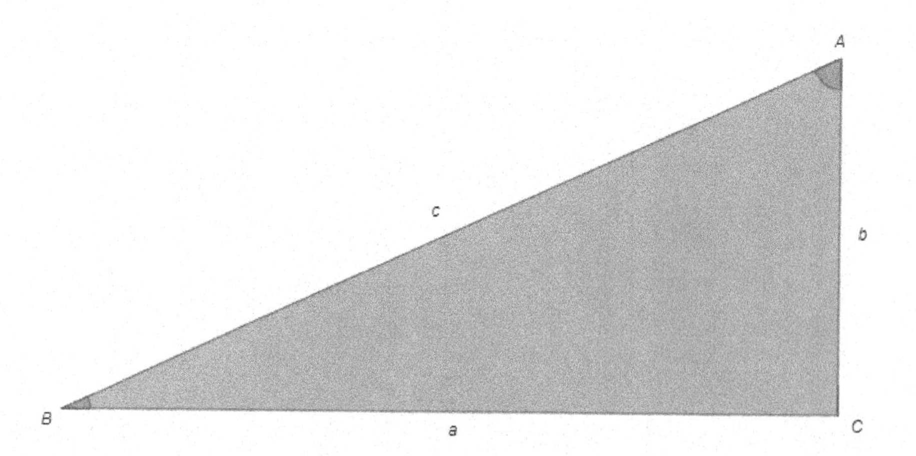

To check your answers above, open the **Guided Visualization** seen on page 9.1-11 of your eText. Input the parameters for the two triangles, above, to check your solutions. You may want to come up with a couple more triangles to solve to make sure that you understand this concept.

Section 9.1 Objective 2 Applications of Right Triangle Trigonometry

What is the **angle of elevation?**

What is the **angle of depression?**

Sketch and label the diagram seen on the first page of this objective.

Work through the video with Example 3 and show all work below.

The angle of elevation to the top of a flagpole measured by a digital protractor is 20° from a point on the ground 90 feet away from its base. Find the height of the flagpole. Round to two decimal places.

Work through the video with Example 4 and show all work below.
At the same instant, two observers 5 miles apart are looking at the same airplane. The angle of elevation (from the ground) of the observer closest to the plane is $71°$. The angle of elevation (from the ground) of the person furthest from the plane is $39°$. Find the altitude of the plane (to the nearest foot) at the instant the two people observe the plane.

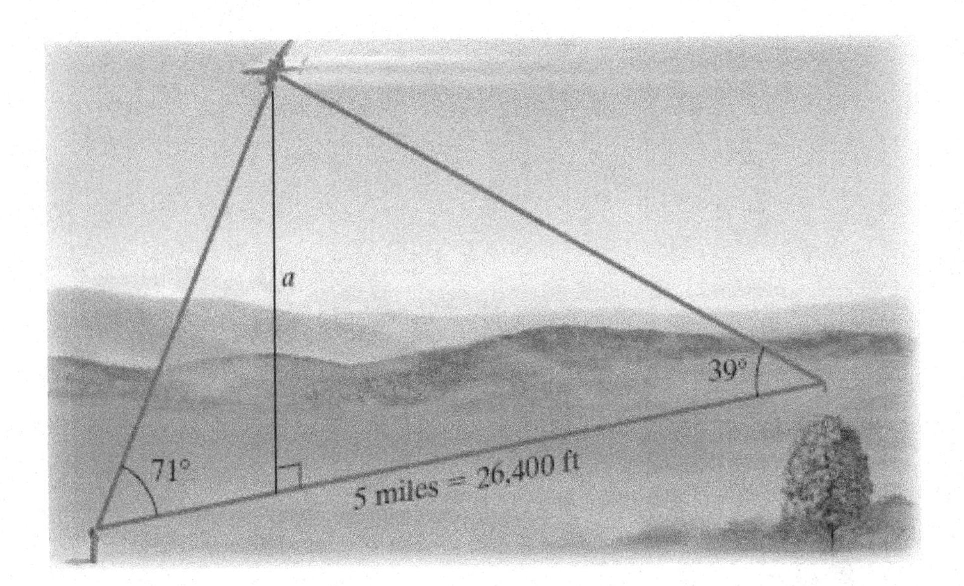

Work through the video with Example 5 and show all work below.
A tourist visiting Paris determines that the angle of elevation from a point A to the top of the Eiffel tower is $12.19°$. She then walks 1 km on a straight line toward the tower to point B and determines that the angle of elevation is to the top of the tower is $32.94°$. Determine the height of the Eiffel Tower. Round to the nearest meter.

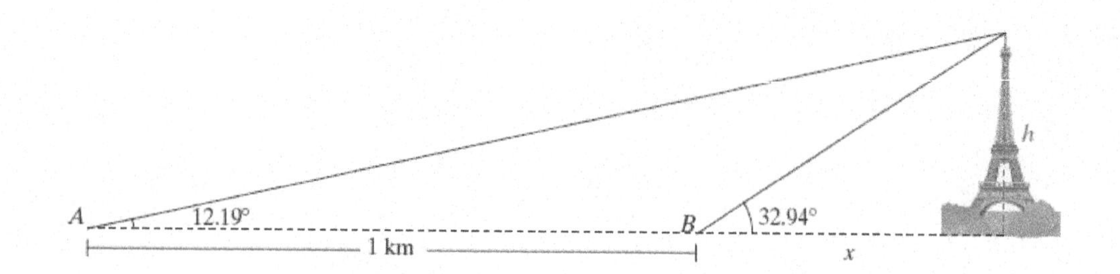

Section 9.2 Guided Notebook

9.2 The Law of Sines

- ▯ Work through Section 9.2 TTK #1
- ▯ Work through Section 9.2 Objective 1
- ▯ Work through Section 9.2 Objective 2
- ▯ Work through Section 9.2 Objective 3
- ▯ Work through Section 9.2 Objective 4

Section 9.2 The Law of Sines

Section 9.2 Objective 1 Determining If the Law of Sines Can Be Used to Solve an Oblique Triangle

What are the two types of **oblique triangles**? (Describe and sketch.)

What is the **Law of Sines**?

Describe the six cases of oblique triangles, as seen in Table 1.

What three pieces of information are needed to solve an oblique triangle using the Law of Sines?

1.

2.

3.

Work through the video with Example 1 showing all work below.
Decide whether or not the Law of Sines can be used to solve each triangle.
Do not attempt to solve the triangle.

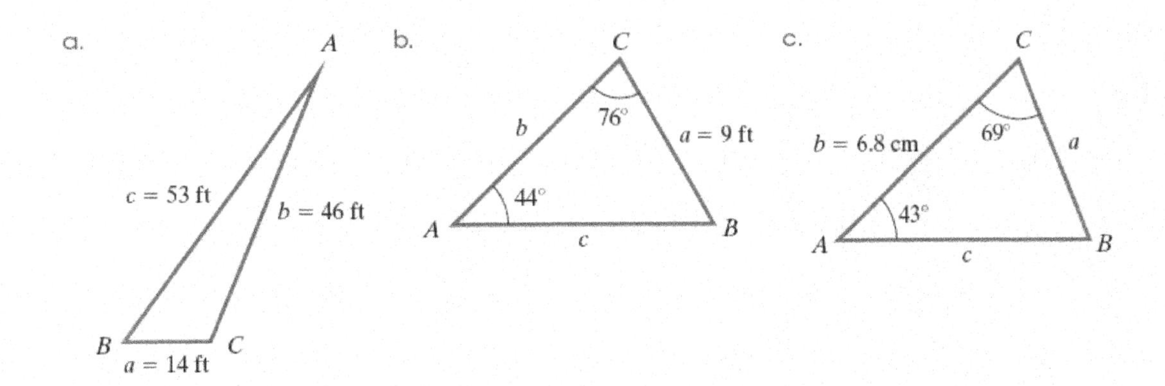

Section 9.2 Objective 2 Using the Law of Sines to Solve the SAA Case or ASA Case

Work through Example 2 showing all work below.
Solve the given oblique triangle. Round the lengths of the sides to one decimal place.

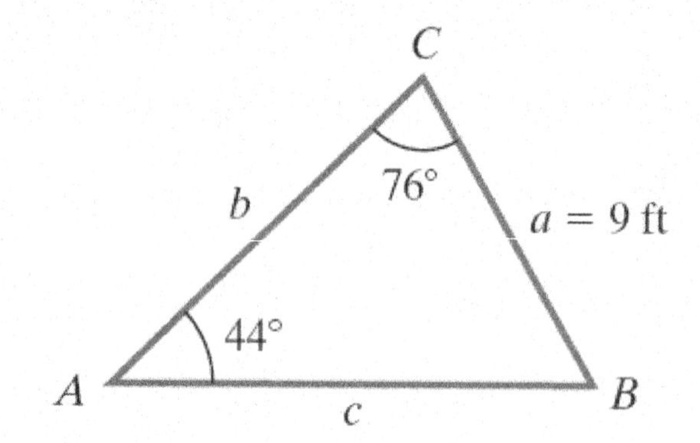

Work through the video with Example 3 showing all work below.
Solve the oblique triangle *ABC* if $B = 38°$, $C = 72°$, and $a = 7.5$ cm.
Round the lengths of the sides to one decimal place.

Section 9.2 Objective 3 Using the Law of Sines to Solve the SSA Case

Watch the animation found in this objective to see all of the possible triangles that can result when given the SSA case and take notes here:

Describe the different scenarios for the SSA case as seen in Table 2.

Work through Example 4 and show all work below.
Two sides and an angle are given below. Determine whether the information results in no triangle, one right triangle, or one or two oblique triangles. Solve each resulting triangle. Round the measures of all angles and the lengths of all sides to one decimal place.

$a = 10$ ft, $b = 28$ ft, $A = 29°$

Work through the video with Example 5 and show all work below.
Two sides and an angle are given below. Determine whether the information results in no triangle, one right triangle, or one or two oblique triangles. Solve each resulting triangle. Round the measures of all angles and the lengths of all sides to one decimal place.

$a = 13$ cm, $b = 7.8$ cm, $A = 67°$

Work through the video with Example 6 and show all work below.
Two sides and an angle are given below. Determine whether the information results in no triangle, one right triangle, or one or two oblique triangles. Solve each resulting triangle. Round the measures of all angles and the lengths of all sides to one decimal place.

$b = 11.3$ in., $c = 15.5$ in., $B = 34.7°$

What are the three cases for which the Law of Sines can be used? (Sketch and label.)

Section 9.2 Objective 4 Using the Law of Sines to Solve Applied Problems Involving Oblique Triangles

Work through Example 7 and show all work below.
To determine the width of a river, forestry workers place markers on opposite sides of the river at points A and B. A third marker is placed at point C, 200 feet away from point A. If the angle between point C and B is $51°$ and if the angle between point A and point C is $110°$, then determine the width of the river rounded to the nearest tenth of a foot.

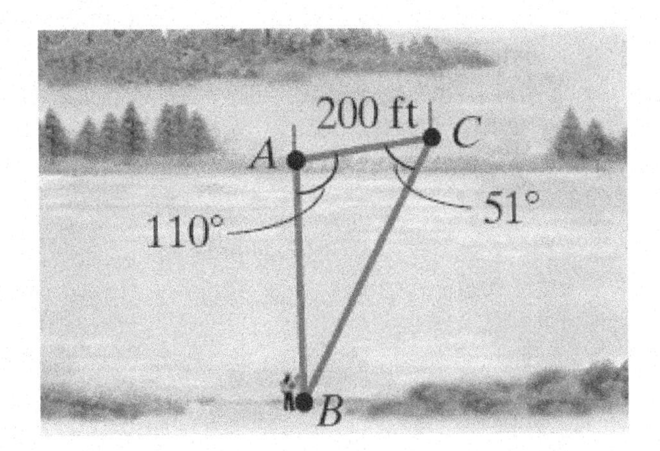

Describe the concept of **bearing**.

Work through the video with Example 8 and show all work below.
A ship set sail from port at a bearing of N 53° E and sailed 63 km to point *B*. The ship then turned and sailed an additional 69 km to point *C*. Determine the distance from port to point *C* if the ship's final bearing is N 74° E. Round to the nearest tenth of a kilometer.

Section 9.3 Guided Notebook

9.3 The Law of Cosines

- ☐ Work through Section 9.3 TTK #1 and 2
- ☐ Work through Section 9.3 Objective 1
- ☐ Work through Section 9.3 Objective 2
- ☐ Work through Section 9.3 Objective 3
- ☐ Work through Section 9.3 Objective 4

Section 9.3 The Law of Cosines

Section 9.3 Objective 1 Determining If the Law of Sines or the Law of Cosines Should Be Used to Begin to Solve an Oblique Triangle

The Law of Sines can be used to solve which three cases of triangles?

What is the **Law of Cosines**?

What is the **Alternate Form of the Law of Cosines**?

Work through the video with Example 1 showing all work below.
Decide whether the Law of Sines or the Law of Cosines should be used to begin to solve the given triangle. Do not solve the triangle.

a.

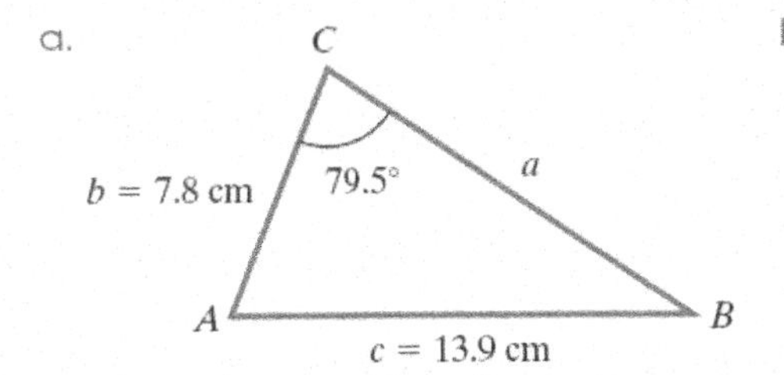

b.

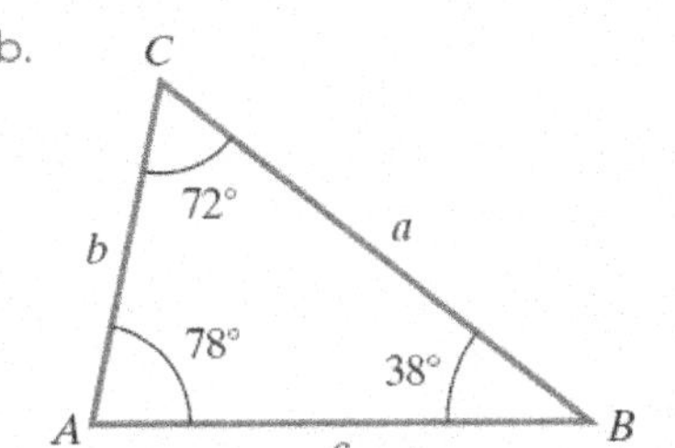

c.

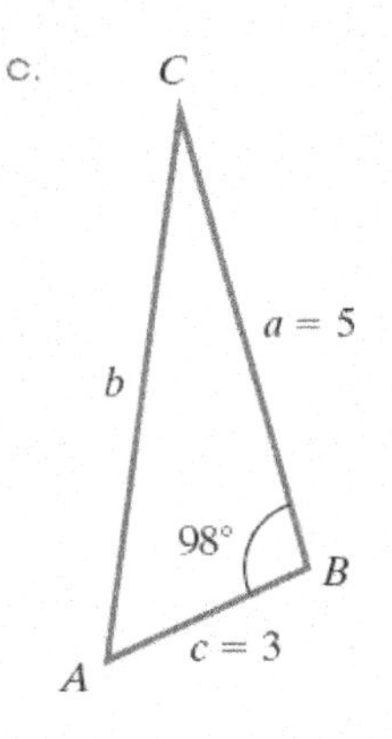

Section 9.3 Objective 2 Using the Law of Cosines to Solve the SAS Case

What are the three steps for **Solving an SAS Oblique Triangle**?

Step 1.

Step 2.

Step 3.

Work through the interactive video with Example 2 showing all work below.
Solve the given oblique triangle. Round the measures of all angles and the lengths of all sides to one decimal place.

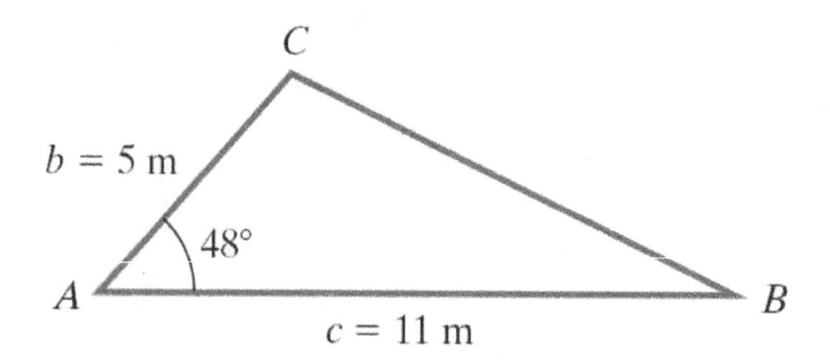

Section 9.3 Objective 3 Using the Law of Cosines to Solve the SSS Case

What are the three steps for **Solving a SSS Oblique Triangle**?

Step 1.

Step 2.

Step 3.

Work through the interactive video with Example 3 and show all work below.
Solve oblique triangle *ABC* if $a = 5$ ft, $b = 8$ ft, and $c = 12$ ft.

Section 9.3 Objective 4 Using the Law of Cosines to Solve Applied Problems Involving Oblique Triangles

Work through the video with Example 4 and show all work below.
Two planes take off from different runways at the same time. One plane flies at an average speed of 350 mph with a bearing of $N\ 21^\circ\ E$. The other plane flies at an average speed of 420 mph with a bearing of $S\ 84^\circ\ W$. How far are the planes from each other 2 hours after takeoff? Round to the nearest tenth of a mile.

Section 10.1 Guided Notebook

10.1 Polar Coordinates and Polar Equations

- ▯ Work through Section 10.1 TTK #1–5
- ▯ Work through Section 10.1 Objective 1
- ▯ Work through Section 10.1 Objective 2
- ▯ Work through Section 10.1 Objective 3
- ▯ Work through Section 10.1 Objective 4
- ▯ Work through Section 10.1 Objective 5
- ▯ Work through Section 10.1 Objective 6

Section 10.1 Polar Coordinates and Polar Equations

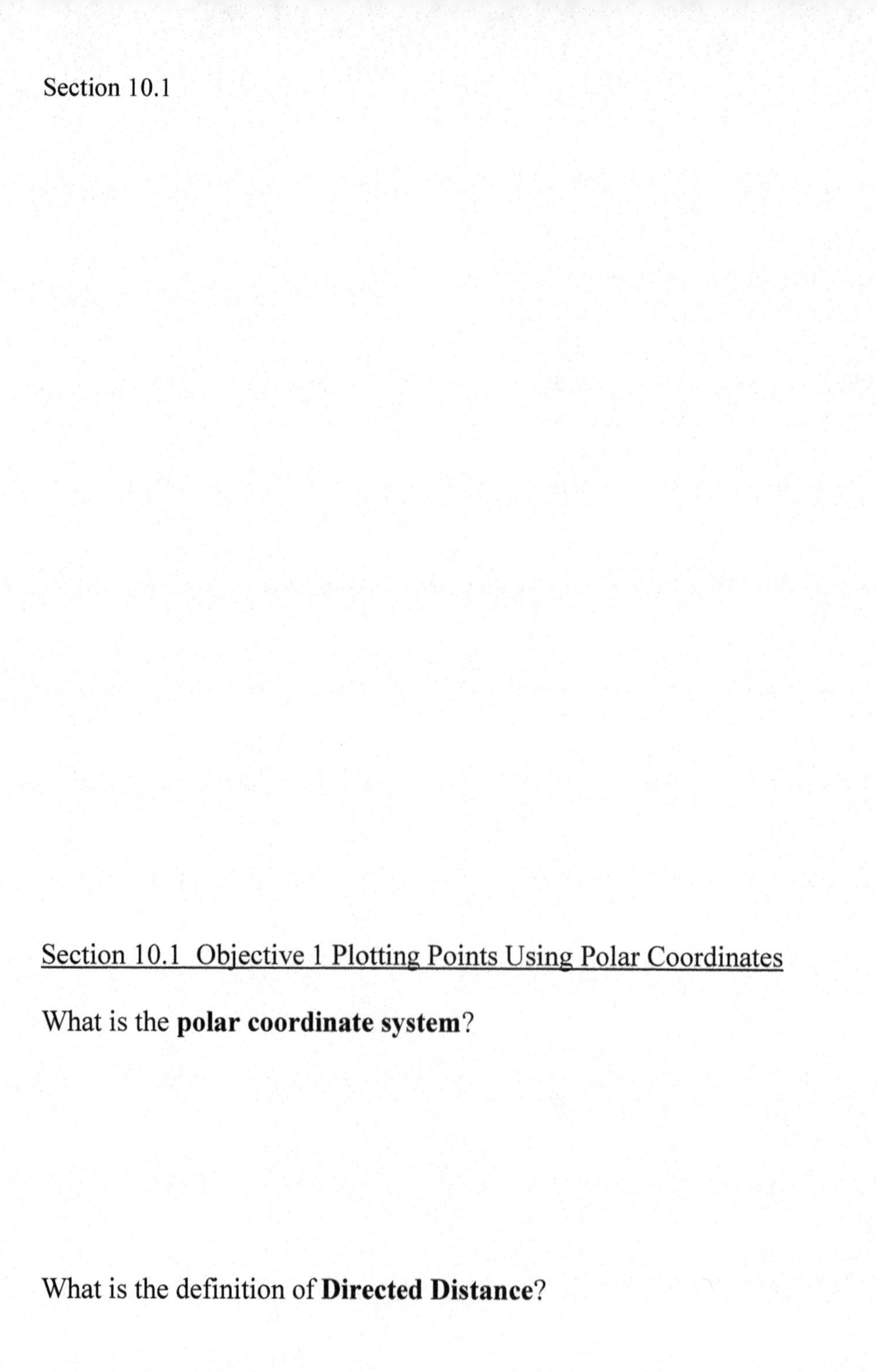

Section 10.1 Objective 1 Plotting Points Using Polar Coordinates

What is the **polar coordinate system**?

What is the definition of **Directed Distance**?

Sketch and label a polar grid.

Work through the video with Example 1 showing all work below.
Plot the following points in a polar coordinate system.

a. $A\left(3, \frac{\pi}{4}\right)$

b. $B(-2, 120°)$

c. $C\left(1.5, -\frac{7\pi}{6}\right)$

d. $D\left(-3, -\frac{3\pi}{4}\right)$

Click on the **Guided Visualization** titled "Plotting Polar Coordinates," which can be found on the bottom of page 10.1-9.

After opening this **Guided Visualization,** plot polar coordinates having the following parameters.

1. Plot and label a point in the polar coordinate system such that $r > 0$ and $\theta < 0$.

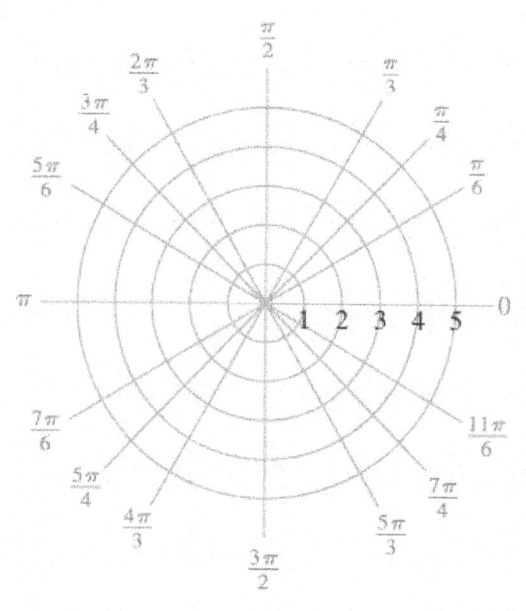

2. Plot and label a point in the polar coordinate system such that $r < 0$ and $\theta > 0$.

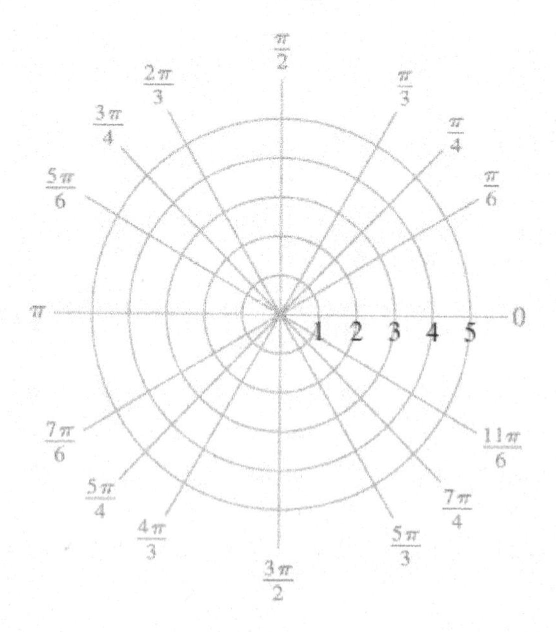

Section 10.1 Objective 2 Determining Different Representations of the Point (r, θ)

In the video that accompanies this objective (this is the same video as seen in Example 1), the points $A\left(3, \frac{\pi}{4}\right)$ and $D\left(-3, -\frac{3\pi}{4}\right)$ are positioned in the exact same location. Plot these two points on the same polar grid below to convince yourself that these two points indeed are positioned at the same location.

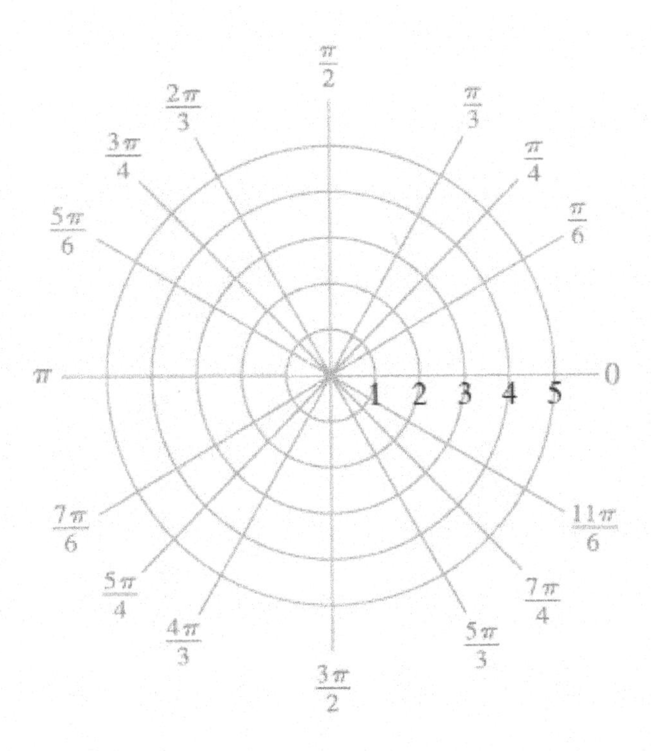

Write down the two ways to determine different representations for a point (r, θ).

Work through the video with Example 2 showing all work below.

The point $P\left(4, \frac{5\pi}{6}\right)$ is shown below:

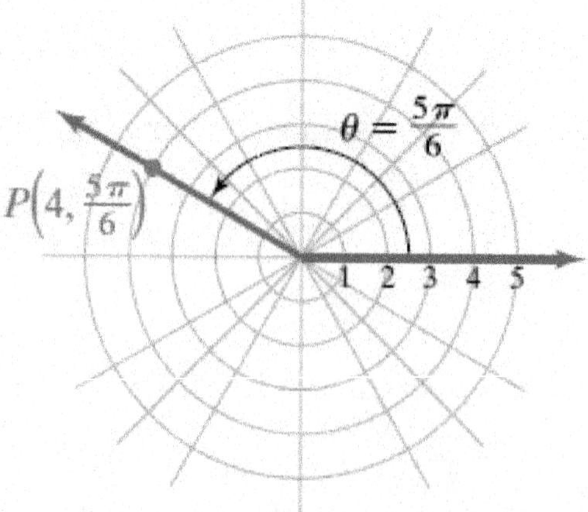

Determine three different representations of point P that have the specified conditions.

a. $r > 0, \ -2\pi \leq \theta < 0$

b. $r < 0, \ 0 \leq \theta < 2\pi$

c. $r > 0, \ 2\pi \leq \theta < 4\pi$

Section 10.1 Objective 3 Converting a Point from Polar Coordinates to Rectangular Coordinates

What are the **Relationships Used when Converting a Point from Polar Coordinates to Rectangular Coordinates**?

Work through the video with Example 3 and show all work below.
Determine the rectangular coordinates for the points with the given polar coordinates.

a. $A(5, \pi)$

b. $B\left(-7, -\frac{\pi}{3}\right)$

c. $C\left(3\sqrt{2}, \frac{5\pi}{4}\right)$

Section 10.1 Objective 4 Converting a Point from Rectangular Coordinates to Polar Coordinates

For simplicity and consistency, we will always determine the polar

coordinates with the conditions that ________________ and ________________.

Work through the video with Example 4 and show all work below.
Determine the polar coordinates for the points with the given rectangular coordinates.

a. $A(-3.5, 0)$

b. $B(0, -\sqrt{7})$

What are the four steps for **Converting Rectangular Coordinates to Polar Coordinates for Points Not Lying Along an Axis**?

Step 1.

Step 2.

Step 3.

Step 4.

Work through the interactive video with Example 5 and show all work below.
Determine the polar coordinates for the points with the given rectangular coordinates such that $r \geq 0$ and $0 \leq \theta < 2\pi$.

a. $A(-4,-4)$

b. $B(-2\sqrt{3},2)$

c. $C(4,-3)$

Section 10.1 Objective 5 Converting an Equation from Rectangular Form to Polar Form

What is a **polar equation**? (Define and give examples.)

Work through the video with Example 6 and show all work below.
Convert each equation given in rectangular form into polar form.

a. $x = 7$

b. $2x - y = 8$

c. $4x^2 + 4y^2 = 3$

d. $x^2 + y^2 = 9y$

Section 10.1 Objective 6 Converting an Equation from Polar Form to Rectangular Form

Work through the video with Example 7 and show all work below.
Convert each equation given in polar form into rectangular form.

a. $3r\cos\theta - 4r\sin\theta = -1$

b. $r = 6\cos\theta$

c. $r = 3$

d. $\theta = \dfrac{\pi}{6}$

Section 10.2 Guided Notebook

10.2 Graphing Polar Equations

- [ ] Work through Section 10.2 TTK #1–4
- [ ] Work through Section 10.2 Objective 1
- [ ] Work through Section 10.2 Objective 2
- [ ] Work through Section 10.2 Objective 3
- [ ] Work through Section 10.2 Objective 4
- [ ] Work through Section 10.2 Objective 5

Section 10.2 Graphing Polar Equations

Section 10.2 Objective 1 Sketching Equations of the Form $\theta = \alpha$, $r\cos\theta = a$, $r\sin\theta = a$, and $ar\cos\theta + br\sin\theta = c$

Work through the video with Example 1 showing all work below.

Sketch the graph of the polar equation $\theta = \frac{2\pi}{3}$.

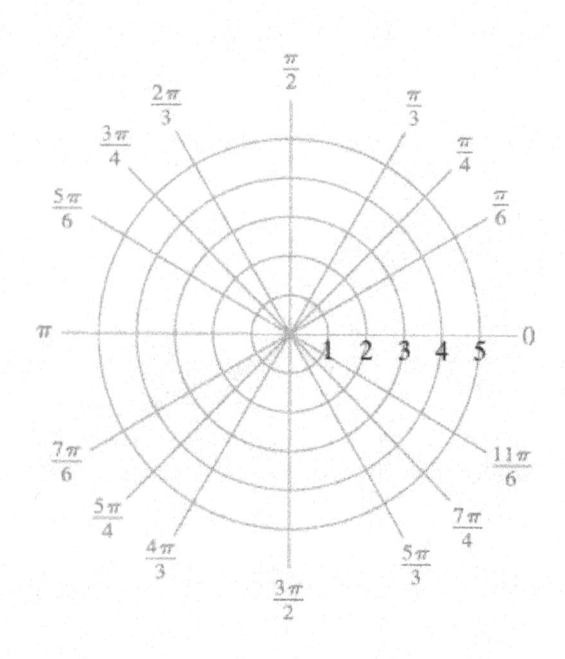

Work through Example 2 showing all work below.
Sketch the graph of the polar equation $r\cos\theta = 2$.

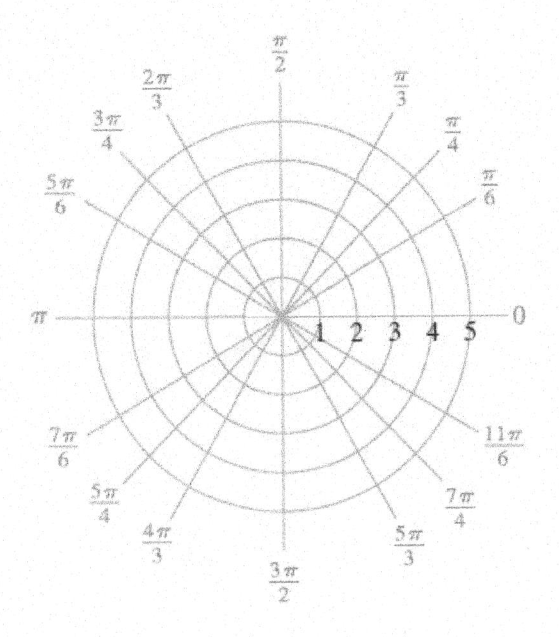

Work through the video with Example 3 showing all work below.
Sketch the graph of the polar equation $r\sin\theta = -3$.

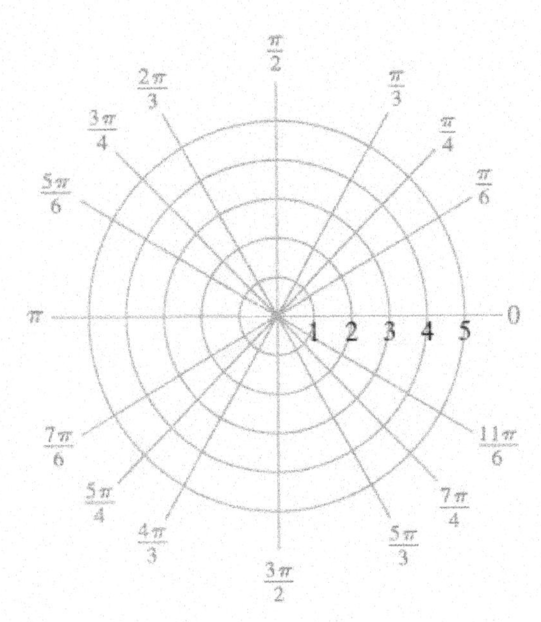

Work through Example 4 showing all work below.
Sketch the graph of the polar equation $3r\cos\theta - 2r\sin\theta = -6$

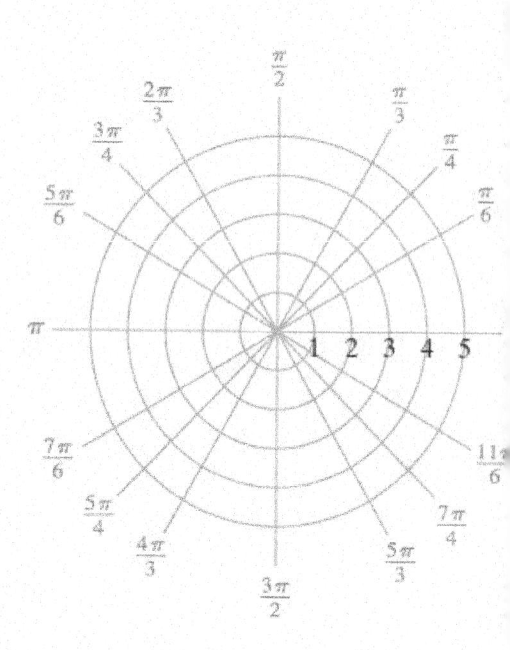

Write down the summary, **Graphs of Polar Equations of the Form** $\boldsymbol{\theta = \alpha}$**,** $\boldsymbol{r\cos\theta = a}$**,** $\boldsymbol{r\sin\theta = a}$**, and** $\boldsymbol{ar\cos\theta + br\sin\theta = c}$**, where** ***a, b,*** **and** ***c*** **are Constants.** (Be sure to sketch each graph.)

Section 10.2 Objective 2 Sketching Equations of the Form $r = a$, $r = a \sin\theta = a$, and $r = a \cos\theta$

Write the polar equation $r = a$ in rectangular form.

Work through Example 5 showing all work below.
Sketch the graph of the polar equation $r = -3$.

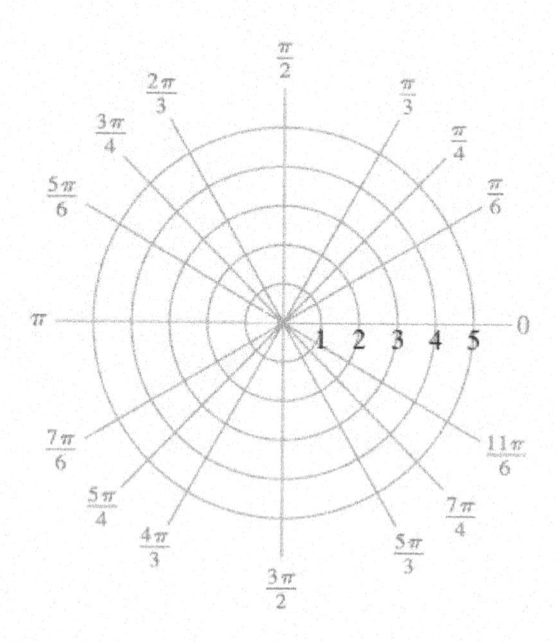

Work through the interactive video with Example 6 showing all work below.
Sketch the graph of each polar equation.

a. $r = 4\sin\theta$

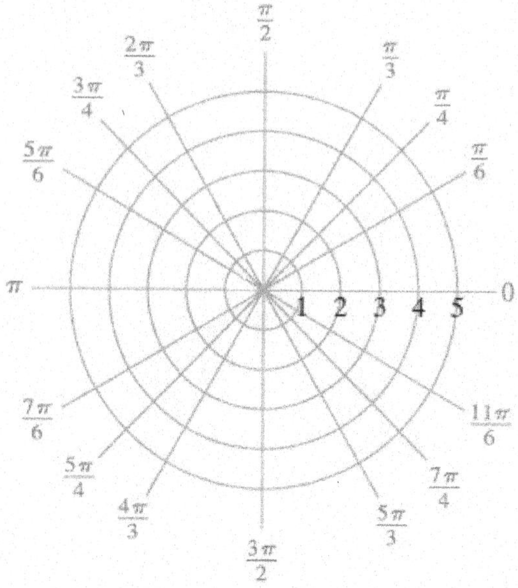

b. $r = -2\cos\theta$

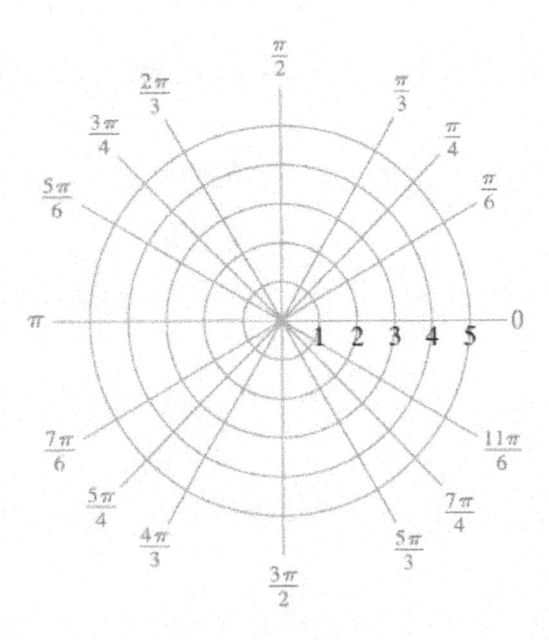

Click on the **Guided Visualization** on the bottom of page 10.2.17. Use this **Guided Visualization** to sketch the following functions.

1. $r = 2$

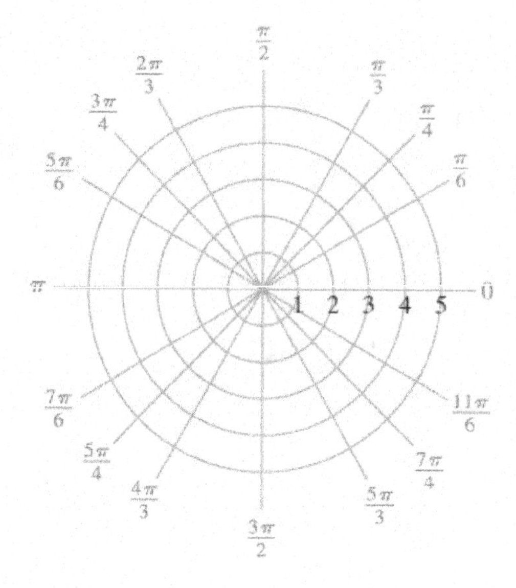

2. $r = -3\sin\theta$

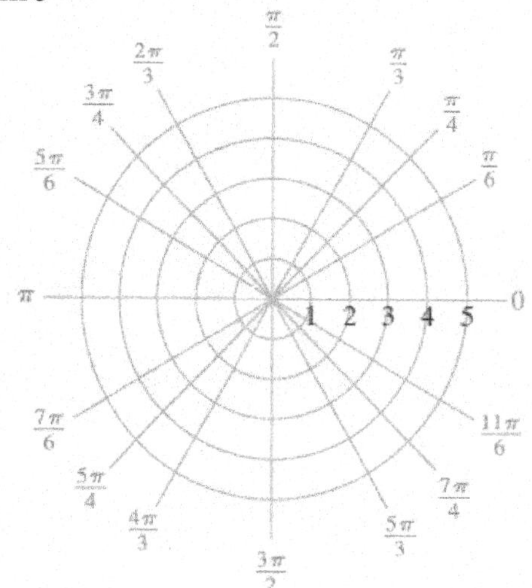

3. $r = 2\cos\theta$

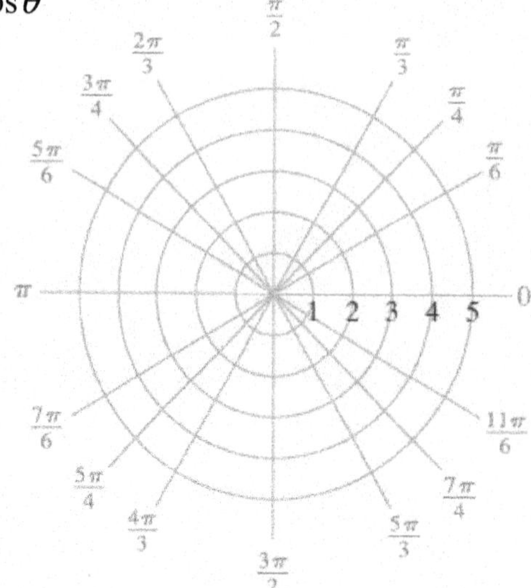

Write down the summary, **Graphs of Polar Equations of the Form** $r = a$, $r = a\sin\theta$, **and** $r = a\cos\theta$, **where** $a \neq 0$ **is a Constant.** (Be sure to sketch the graph of each.)

Section 10.2 Objective 3 Sketching Equations of the Form $r = a + b\sin\theta$ and $r = a + b\cos\theta$

What are **limacons**? What form do their equations have?

Work through Example 7 showing all work below.
Sketch the graph of the polar equation $r = 3 - 3\sin\theta$.

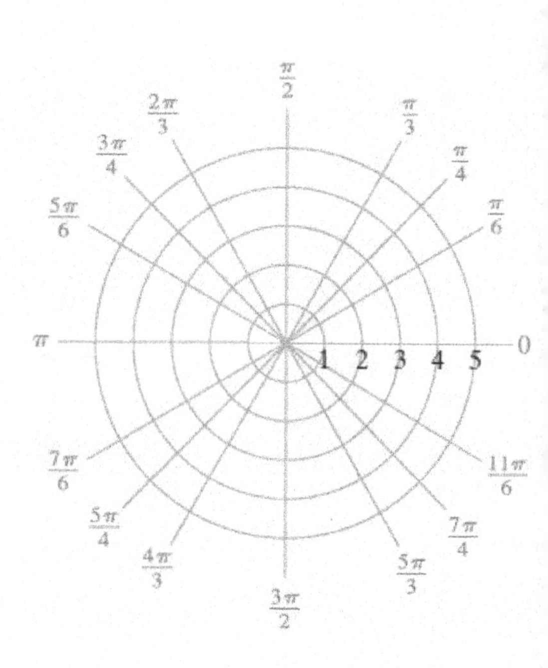

What is a **cardioid**?

Work through Example 8 showing all work below.
Sketch the graph of the polar equation $r = -1 + 2\cos\theta$.

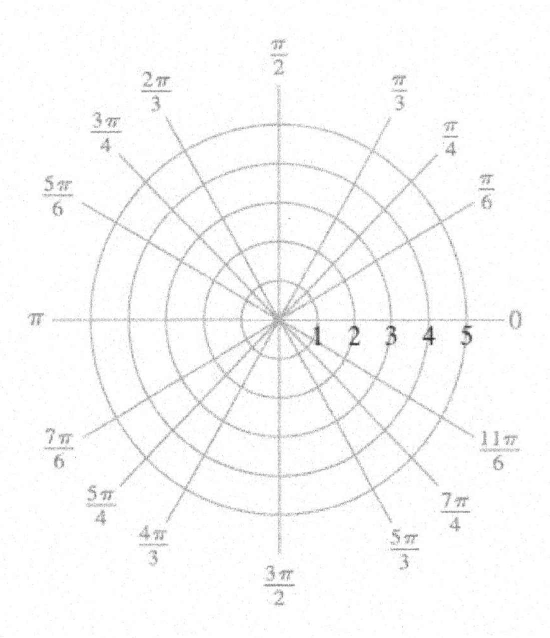

The graph from Example 8 is an example of a limacon that has

an ________________________________.

Work through Example 9 showing all work below.
Sketch the graph of the polar equation $r = 3 + 2\sin\theta$.

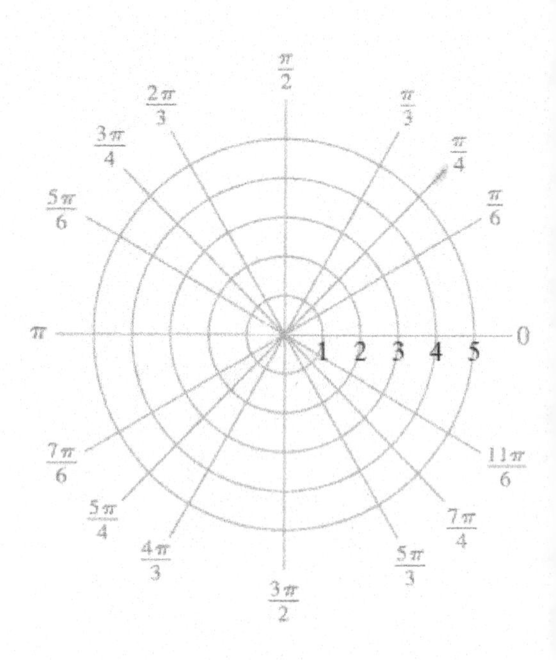

The graph from Example 9 is an example of a limacon that has

an________________________________.

Write down the summary, **Graphs of Polar Equations of the Form $r = a + b\sin\theta$, and $r = a + b\cos\theta$, where $a \neq 0$ and $b \neq 0$ Are Constants.** (Be sure to sketch **one graph** of each type.)

What are the four **Steps for Sketching Polar Equations (Limacons) of the Form $r = a + b\sin\theta$ and $r = a + b\cos\theta$?**

Step 1.

Step 2.

Step 3.

Step 4.

Click on the **Guided Visualization** on page 10.2.31. See if you can first sketch the graph of $r = 1 + 2\cos\theta$ using the four-step process outlined on the previous page. Then, use this **Guided Visualization** on page 10.2-31 to check your answer.

Sketch the graph of $r = 1 + 2\cos\theta$.

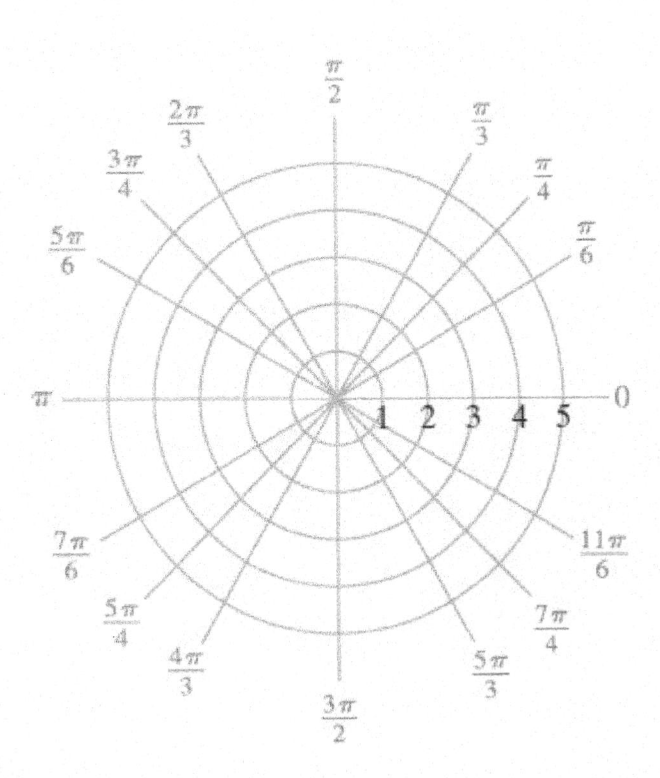

Work through the interactive video with Example 10 showing all work below.
Sketch the graph of each polar equation.

a. $r = 4 - 3\cos\theta$

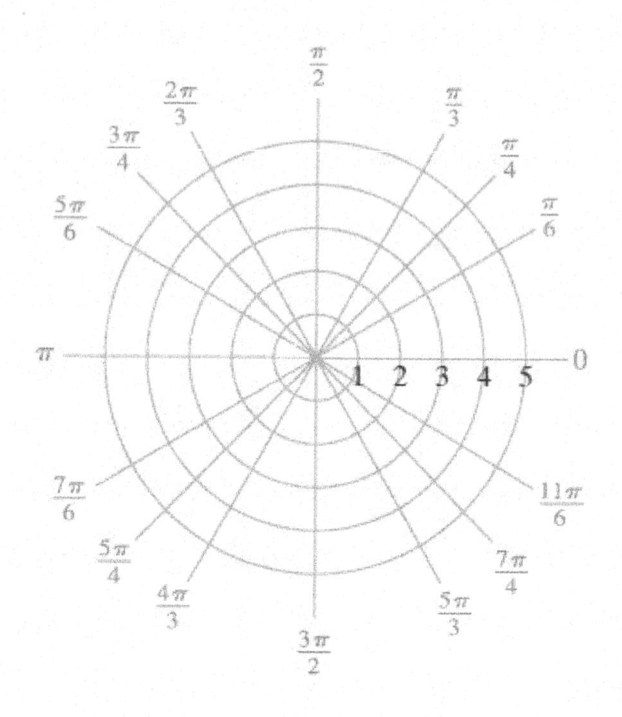

b. $r = 2 + \sin\theta$

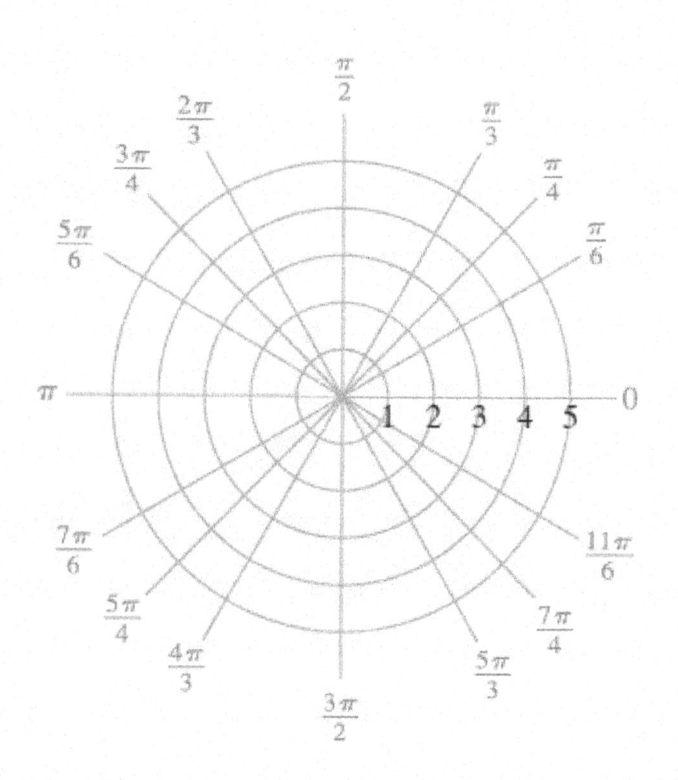

c. $r = -2 + 2\cos\theta$

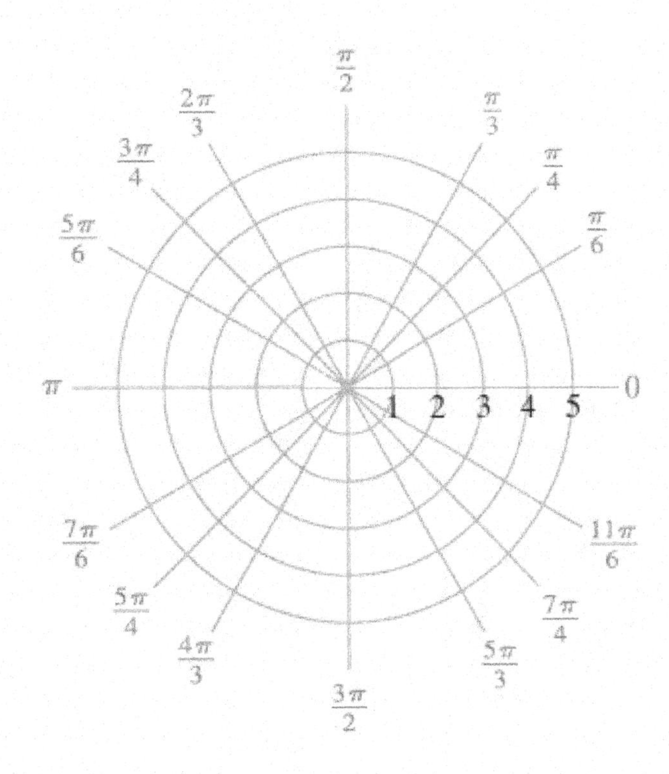

d. $r = 3 - 4\sin\theta$

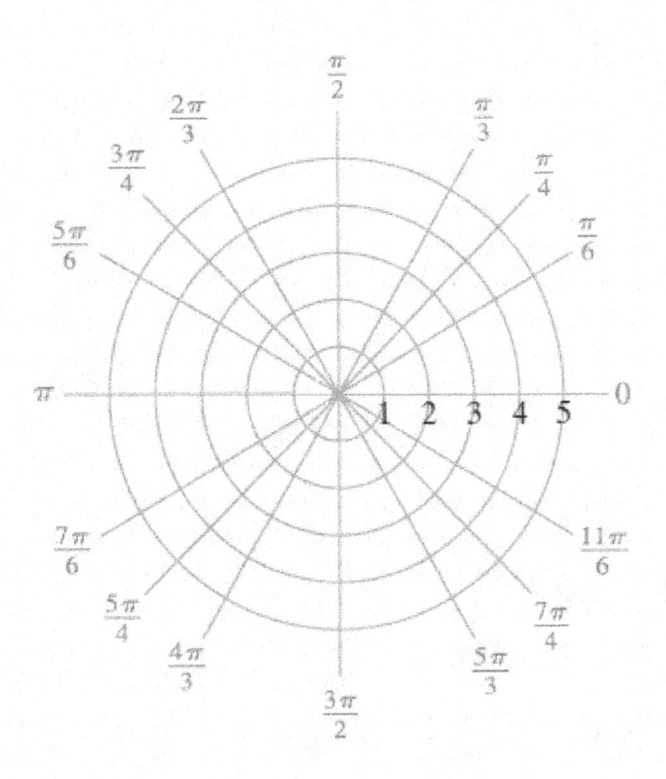

Section 10.2 Objective 4 Sketching Equations of the Form $r = a \sin n\theta$ and $r = a \cos n\theta$

What do we call the graphs of these types of equations? Why?

Work through Example 11 and show all work below.
Sketch the graph of $r = 3\sin 2\theta$.

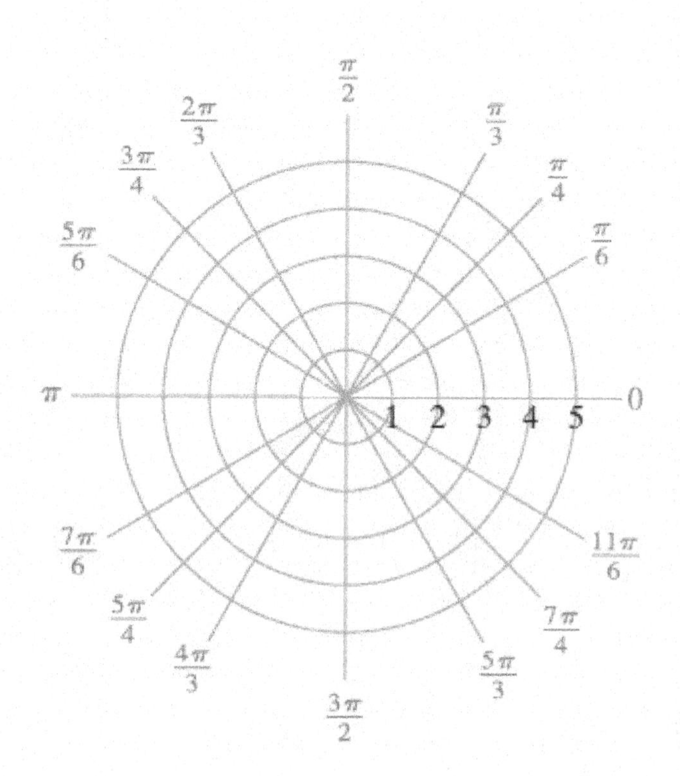

In the equations $r = a\sin n\theta$ or $r = a\cos n\theta$,

if n is **even**, there will be ______________ petals.

if n is **odd**, there will be ______________ petals.

Write down the summary, **Graphs of Polar Equations of the Form $r = a\sin n\theta$, and $r = a\cos n\theta$, where $a \neq 0$ Is a Constant and $n \neq 1$ Is a Positive Integer.**
(Be sure to sketch a graph of each type.)

What are the six **Steps for Sketching Polar Equations (Roses) of the Form $r = a \sin n\theta$, and $r = a \cos n\theta$, where $a \neq 0$ Is a Constant and $n \neq 1$ Is a Positive Integer?**

Step 1.

Step 2.

Step 3.

Step 4.

Step 5.

Step 6.

Click on the **Guided Visualization** on the bottom of page 10.2.40. See if you can first sketch the graph of $r = -2\sin 3\theta$ using the six-step process outlined on the previous page. Then, use this **Guided Visualization** on page 10.2-40 to check your answer.

Sketch the graph of $r = -2\sin 3\theta$.

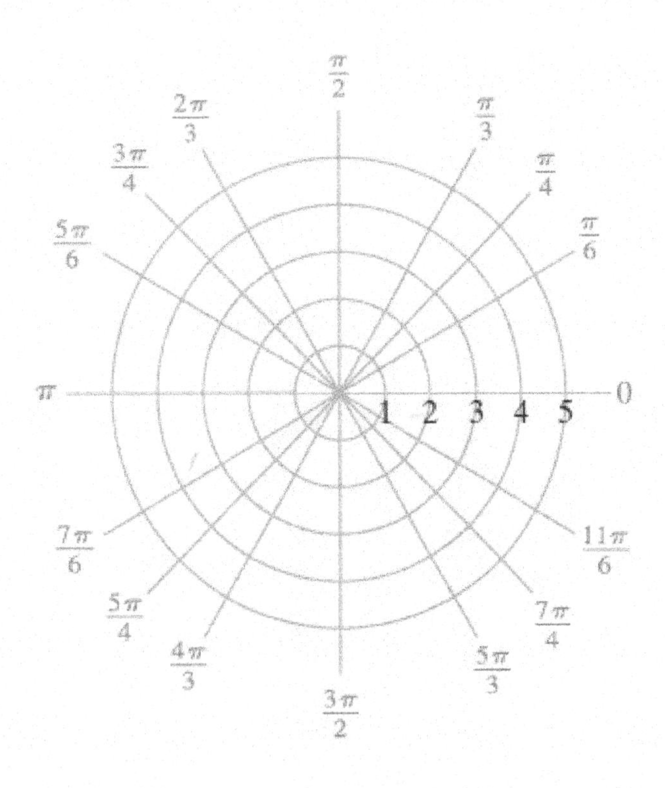

Work through the interactive video with Example 12 showing all work below.
Sketch the graph of each polar equation.

a. $r = -4\cos 3\theta$

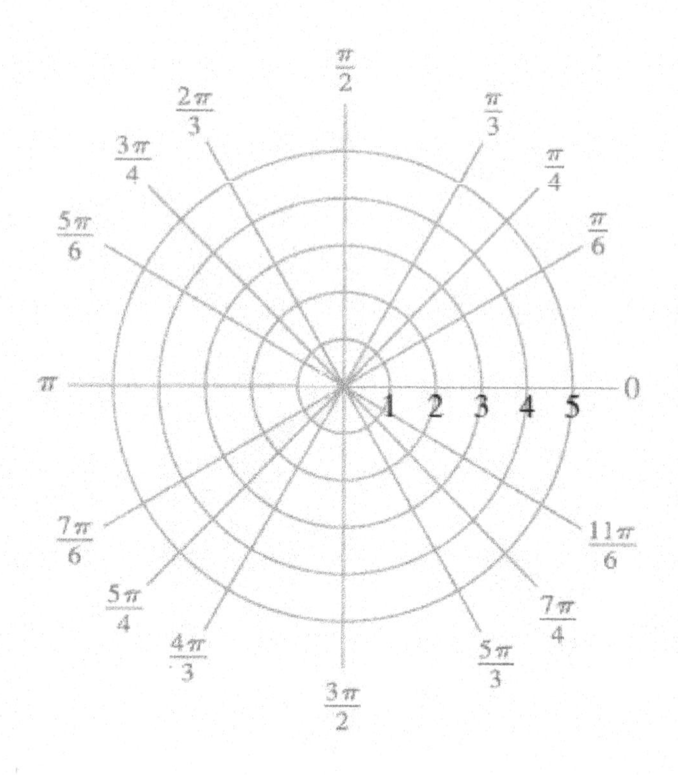

b. $r = -2\sin 5\theta$

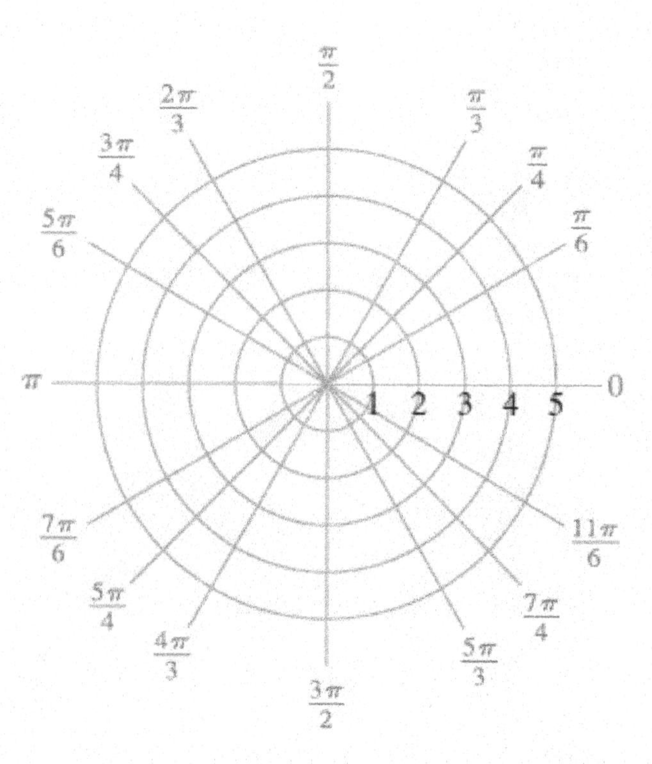

c. $r = 5\cos 4\theta$

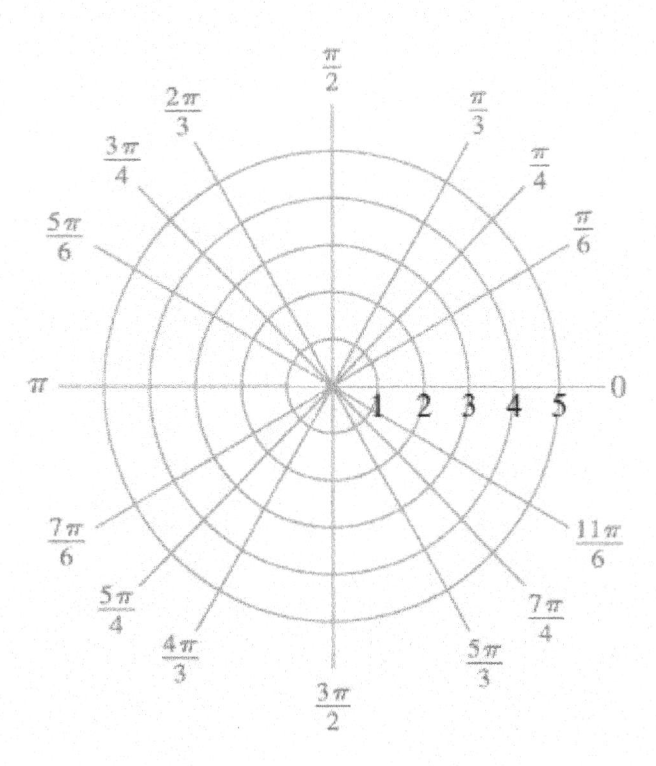

Section 10.3 Guided Notebook

10.3 Complex Numbers in Polar Form; De Moivre's Theorem

- ☐ Work through Section 10.3 TTK #3–6
- ☐ Work through Section 10.3 Objective 1
- ☐ Work through Section 10.3 Objective 2
- ☐ Work through Section 10.3 Objective 3
- ☐ Work through Section 10.3 Objective 4
- ☐ Work through Section 10.3 Objective 5
- ☐ Work through Section 10.3 Objective 6
- ☐ Work through Section 10.3 Objective 7

Section 10.3 Complex Numbers in Polar Form; De Moivre's Theorem

Section 10.3 Objective 1 Understanding the Rectangular Form of a Complex Number

What is the definition of the **Rectangular Form of a Complex Number**?

What is the **complex plane**? (Define and sketch.)

What is the definition of the **Absolute Value of a Complex Number**?

Work through the video with Example 1 showing all work below.
Plot each complex number in the complex plane and determine its absolute value.

a. $z_1 = 3 - 4i$

b. $z_2 = -2 + 5i$

c. $z_3 = 3$

d. $z_4 = -2i$

Section 10.3 Objective 2 Understanding the Polar Form of a Complex Number

What is the definition of the **Polar Form of a Complex Number**?

What is the **modulus**? What is the **argument**?

Complex numbers in polar form, where $0 \leq \theta < 2\pi$ (or $0 \leq \theta < 360°$) are said

to be in ________________________________.

Work through the video with Example 2 showing all work below.
Rewrite the complex number in standard polar form, plot the number in the complex plane, and determine the quadrant in which the point lies or the axis on which the point lies.

a. $z=3\left(\cos\frac{5\pi}{8}+i\sin\frac{5\pi}{8}\right)$

b. $z=2\left(\cos\frac{23\pi}{4}+i\sin\frac{23\pi}{4}\right)$

c. $z=4(\cos(-3\pi)+i\sin(-3\pi))$

Section 10.3 Objective 3 Converting a Complex Number from Polar Form to Rectangular Form

Work through the video with Example 3 showing all work below.
Write each complex number in rectangular form using exact values if possible. Otherwise, round to two decimal places.

a. $z = 3\left(\cos\frac{7\pi}{4} + i\sin\frac{7\pi}{4}\right)$

b. $z = 4(\cos 80° + i\sin 80°)$

Section 10.3 Objective 4 Converting a Complex Number from Rectangular Form to Standard Polar Form

Write down the four cases outlined in **Converting Complex Numbers From Rectangular Form to Standard Polar Form for Complex Numbers Lying Along the Real Axis or Imaginary Axis**.

Work through the video with Example 4 showing all work below.
Determine the standard polar form of each complex number. Write the argument using radians.

a. $z = 5$

b. $z = -3i$

c. $z = -\sqrt{7}$

d. $z = \frac{7}{2}i$

What are the four steps for **Converting a Complex Number from Rectangular Form to Standard Polar Form for** $a \neq 0$ **and** $b \neq 0$ **?**

Step 1.

Step 2.

Step 3.

Step 4.

Work through the interactive video with Example 5 showing all work below.
Determine the standard polar form of each complex number. Write the argument in radians using exact values if possible. Otherwise, round the argument to two decimal places.

a. $z = -2\sqrt{3} + 2i$

b. $z = 4 - 3i$

Section 10.3 Objective 5 Determining the Product or Quotient of Complex Numbers in Polar Form

How do we compute the **Product and Quotient of Two Complex Numbers Written in Polar Form**?

Work through the video with Example 6 showing all work below.

Let $z_1 = 4\left(\cos\frac{2\pi}{3} + i\sin\frac{2\pi}{3}\right)$ and $z_2 = 5\left(\cos\frac{11\pi}{6} + i\sin\frac{11\pi}{6}\right)$. Find $z_1 z_2$ and $\frac{z_1}{z_2}$ and write the answers in standard polar form.

Section 10.3 Objective 6 Using De Moivre's Theorem to Raise a Complex Number to a Power

If $z = r(\cos\theta + i\sin\theta)$, then use the product of two complex numbers formula to find z^2.

Now, find z^3.

Now, find z^4.

What is **De Moivre's Theorem for Finding Powers of a Complex Number**?

Work through the interactive video with Example 7 showing all work below.

a. Find $\left[5\left(\cos\frac{3\pi}{4}+i\sin\frac{3\pi}{4}\right)\right]^3$ and write your answer in standard polar form.

b. Find $(\sqrt{3}-i)^4$ and write your answer in rectangular form.

Section 11.1 Guided Notebook

Section 11.1 The Parabola

- ☐ Work through Objective 1
- ☐ Work through Objective 2
- ☐ Work through Objective 3
- ☐ Work through Objective 4
- ☐ Work through Objective 5

Introduction to Conic Sections

How are conic sections formed? Draw a picture below.

What are the four conic sections that can be formed?

View the animations for the circle and take notes below.

View the animation for the ellipse and take notes below.

View the animation for the hyperbola and take notes below.

What are degenerate conic sections?

What are the three degenerate conic sections?

Section 11.1 Objective 1: Determining the Equation of a Parabola with a Vertical Axis of Symmetry

View the animation and describe the characteristics of a parabola.

What is the **geometric definition of a parabola**?

Watch the video seen just after the geometric definition of the parabola and sketch a parabola using this geometric definition below.

Equation of a Parabola in Standard Form with a Vertical Axis of Symmetry

The equation of a parabola with a vertical axis of symmetry is

______________________________________, where

the vertex is _____,

______ = distance from vertex to the focus or the distance from the vertex to the directrix,

the focus is ___________________, and

the equation of the directrix is ___________________.

Draw two different parabolas having a vertical axis of symmetry below and label the vertex, focus, and directrix.

Work through Example 1 taking notes below.
Find the vertex, focus and directrix of the parabola $x^2 = 8y$ and sketch its graph.

Work through the video for Example 2 and take notes below.
Find the vertex, focus, and directrix of the parabola $-(x+1)^2 = 4(y-3)$ and sketch its graph.

Section 11.1 Objective 2: Determining the Equation of a Parabola with a Horizontal Axis of Symmetry

Equation of a Parabola in Standard Form with a Horizontal Axis of Symmetry

The equation of a parabola with a horizontal axis of symmetry is

____, where

the vertex is _____,

______ = distance from vertex to the focus or the distance from the vertex to the directrix,

the focus is ____________________, and

the equation of the directrix is ____________________.

Draw two different parabolas having a horizontal axis of symmetry below and label the vertex, focus, and directrix.

Work through the video with Example 3 and take notes below.

Find the vertex, focus, and directrix of the parabola $(y-3)^2 = 8(x+2)$ and sketch its graph.

Section 11.1 Objective 3: Determining the Equation of a Parabola Given Information about the Graph

Work through the video with Example 4 and take notes below.
Find the standard form of the equation of the parabola with focus $\left(-3, \frac{5}{2}\right)$ and directrix $y = \frac{11}{2}$.

Work through the video with Example 5 and take notes below.
Find the standard form of the equation of the parabola with focus $(4,-2)$ and vertex $\left(\frac{13}{2},-2\right)$.

Section 11.1 Objective 4: Completing the Square to Find the Equation of a Parabola in Standard Form

Work through the video with Example 6 and take notes below.

Find the vertex, focus, and directrix and sketch the graph of the parabola $x^2-8x+12y=-52$.

Section 11.1 Objective 5: Solving Applied Problems Involving Parabolas

What did the Romans create using properties of parabolic structures?

What are some of the objects that are manufactured that use parabolic surfaces?

Work through Example 7 and take notes below.
Parabolic microphones can be seen on the sidelines of professional sporting events so that television networks can capture audio sounds from the players on the field. If the surface of a parabolic microphone is 27 centimeters deep and has a diameter of 72 centimeters at the top, where should the microphone be placed relative to the vertex of the parabola?

Section 11.2 Guided Notebook

Section 11.2 The Ellipse

- ☐ Work through Objective 1
- ☐ Work through Objective 2
- ☐ Work through Objective 3
- ☐ Work through Objective 4

Section 11.2 The Ellipse

View the animation for the ellipse and write the geometric definition of the ellipse. Draw a sketch below demonstrating the definition.

Section 11.2 Objective 1: Sketching the Graph of an Ellipse

Work through the video describing horizontal and vertical ellipses. Draw a sketch and label the key features of each below.

Equation of an Ellipse in Standard Form with Center (h,k)

Write the equation of an ellipse in standard form having a **Horizontal Major Axis:**

What is the relationship between a and b?

What are the coordinates of the foci?

What are the coordinates of the vertices?

What are the endpoints of the major axis?

What is the relationship between a, b, and c?

Draw an ellipse having a horizontal major axis and label the center, foci, endpoints of major axis, and endpoints of minor axis.

Write the equation of an ellipse in standard form having a **Vertical Major Axis:**

What is the relationship between a and b?

What are the coordinates of the foci?

What are the coordinates of the vertices?

What are the endpoints of the major axis?

What is the relationship between a, b, and c?

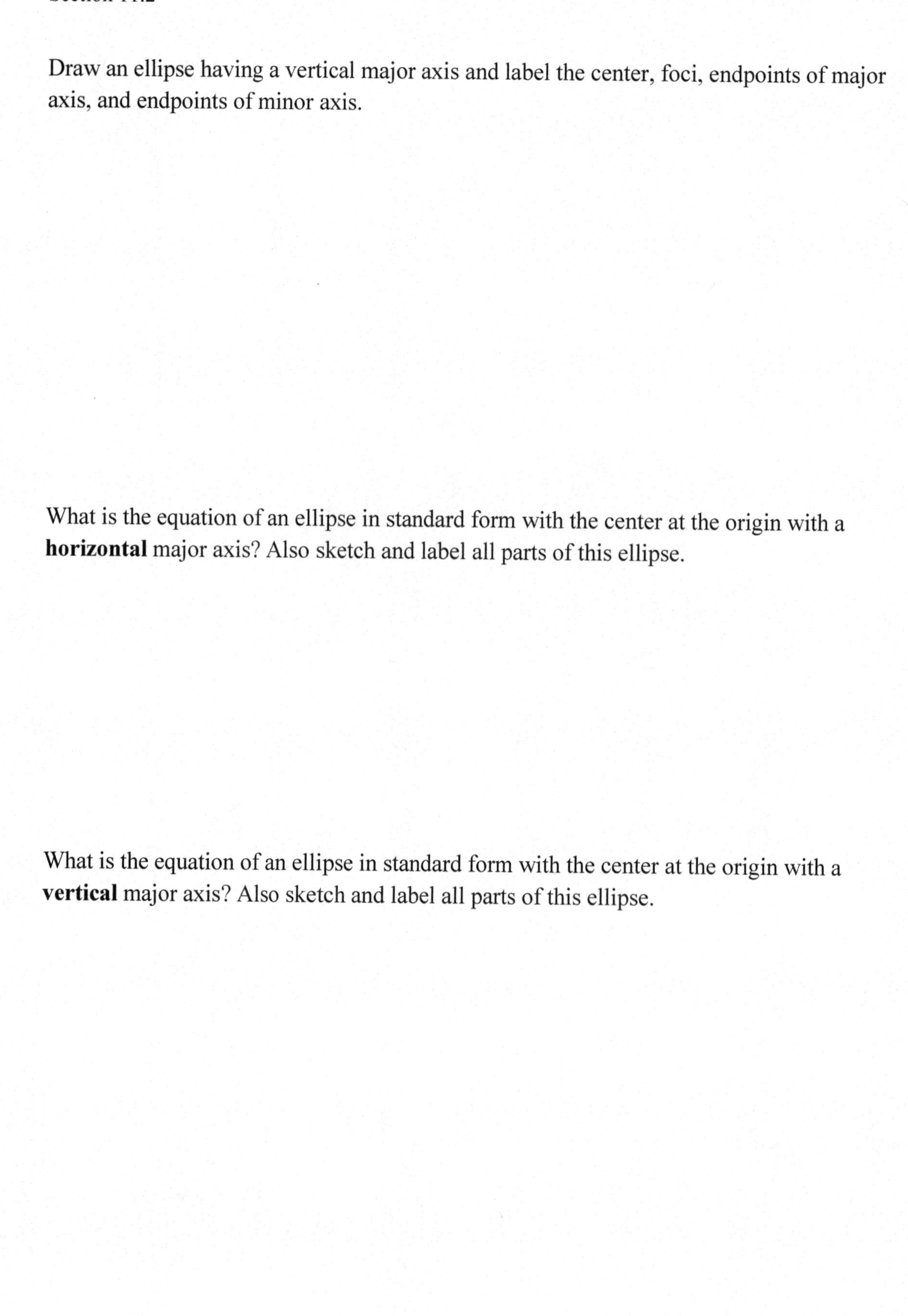

Draw an ellipse having a vertical major axis and label the center, foci, endpoints of major axis, and endpoints of minor axis.

What is the equation of an ellipse in standard form with the center at the origin with a **horizontal** major axis? Also sketch and label all parts of this ellipse.

What is the equation of an ellipse in standard form with the center at the origin with a **vertical** major axis? Also sketch and label all parts of this ellipse.

Work through the video with Example 1 and take notes below.

Sketch the graph of the ellipse $\frac{x^2}{25}+\frac{y^2}{4}=1$, and label the center, foci, and vertices.

Work through the video with Example 2 and take notes below.

Sketch the graph of the ellipse $\frac{(x+2)^2}{20}+\frac{(y-3)^2}{36}=1$, and label the center, foci, and vertices.

Section 11.2 Objective 2: Determining the Equation of an Ellipse Given Information about the Graph

Work through the video with Example 3 and take notes below.
Find the standard form of the equation of the ellipse with foci at $(-6,1)$ and $(-2,1)$ such that the length of the major axis is eight units.

Work through the video with Example 4 and take notes below.
Determine the equation of the ellipse with foci located at $(0,6)$ and $(0,-6)$ that passes through the point $(-5,6)$.

Section 11.2 Objective 3: Completing the Square to Find the Equation of an Ellipse in Standard Form

Work through the video with Example 5 and take notes below.
Find the center and foci and sketch the ellipse $36x^2 + 20y^2 + 144x - 120y - 396 = 0$.

Section 11.2 Objective 4: Solving Applied Problems Involving Ellipses

What is the first application mentioned in the text as an application of ellipses?

What is sound wave lithotripsy?

Work through Example 6 and take notes below.
A patient is placed in an elliptical tank that is 200 cm long and 80 cm wide to undergo sound wave lithotripsy treatment for kidney stones. Determine where the sound emitter and the stone should be positioned relative to the center of the ellipse.

Section 11.3 Guided Notebook

Section 11.3 The Hyperbola

- ☐ Work through Objective 1
- ☐ Work through Objective 2
- ☐ Work through Objective 3
- ☐ Work through Objective 4

Section 11.3 The Hyperbola

View the animation for a hyperbola and take notes below.

What is the geometric definition of a hyperbola?

Section 11.3 Objective 1: Sketching the Graph of a Hyperbola

Work through the video accompanying Objective 1 and describe each of the following:

Center:

Vertices:

Foci:

Transverse axis:

Conjugate Axis:

Asymptotes:

Reference rectangle:

Draw a sketch of two different hyperbolas below labeling these items.

Write down **Fact 1 for Hyperbolas**.

What is the equation of a **Hyperbola in Standard form with a Horizontal Transverse Axis**?

What are the ordered pairs of the foci?

What are the ordered pairs of the vertices?

What are the ordered pairs of the endpoints?

What is the relationship between a, b, and c?

What is the equation of the asymptotes?

What is the equation of a **Hyperbola in Standard form with a Vertical Transverse Axis**?

What are the ordered pairs of the foci?

What are the ordered pairs of the vertices?

What are the ordered pairs of the endpoints?

What is the relationship between a, b, and c?

What is the equation of the asymptotes?

Work through the video with Example 1 and take notes below.
Sketch the following hyperbolas. Determine the center, transverse axis, vertices, and foci and find the equations of the asymptotes.

a. $\frac{(y-4)^2}{36}-\frac{(x+5)^2}{9}=1$

b. $25x^2-16y^2=400$

Section 11.3 Objective 2: Determining the Equation of a Hyperbola in Standard Form

Work through the video with Example 2 and take notes below.
Find the equation of the hyperbola with the center at $(-1,0)$,
a focus at $(-11,0)$, and a vertex at $(5,0)$.

Section 11.3 Objective 3: Completing the Square to Find the Equation of a Hyperbola in Standard Form

Work through the video with Example 3 and take notes below.
Find the center, vertices, foci, and equations of asymptotes and sketch the hyperbola

$12x^2 - 4y^2 - 72x - 16y + 140 = 0$.

Section 11.3 Objective 4: Solving Applied Problems Involving Hyperbolas

List 3 examples of applications of hyperbolas.

1.

2.

3.

Work through Example 4 and take notes below.
One transmitting station is located 100 miles due east from another transmitting station. Each station simultaneously sends out a radio signal. The signal from the west tower is received by a ship $\frac{1,600}{3}$ microseconds after the signal from the east tower. If the radio signal travels at 0.18 miles per microsecond, find the equation of the hyperbola on which the ship is presently located.

Section 12.1 Guided Notebook

Section 12.1 Solving Systems of Linear Equations in Two Variables

- ☐ Work through TTK #1
- ☐ Work through TTK #2
- ☐ Work through TTK #3
- ☐ Work through TTK #4
- ☐ Work through Objective 1
- ☐ Work through Objective 2
- ☐ Work through Objective 3
- ☐ Work through Objective 4

Section 12.1 Solving Systems of Linear Equations in Two Variables

Section 12.1 Objective 1 Verifying Solutions to a System of Linear Equations in Two Variables

Write down the definition of a linear equation in n variables.

What is the important thing to remember about the variables of a linear equation?

Work through Example 1 and take notes below.
Show that the ordered pair $(-1,3)$ is a solution to the system

$$\begin{aligned} 3x - 2y &= -9 \\ x + \quad y &= 2 \end{aligned}.$$

What is a consistent system? Draw a picture of a consistent system.

What is an inconsistent system? Draw a picture of an inconsistent system.

Section 12.1 Objective 2 Solving a System of Linear Equations Using the Substitution Method

Watch the video on substitution and take notes below.

What are the four steps for **Solving a System of Equations by the Method of Substitution?**

Step 1.

Step 2.

Step 3.

Step 4.

Work through Example 2 and take notes below.

Solve the following system using the method of substitution:

$$\begin{aligned} 2x - 3y &= -5 \\ x + y &= 5 \end{aligned}$$

Section 12.1 Objective 3 Solving a System of Linear Equations Using the Elimination Method

Watch the video on the elimination method and take notes below.

What are the five steps for **Solving a System of Equations by the Method of Elimination?**

Step 1.

Step 2.

Step 3.

Step 4.

Step 5.

Work through Example 3 and take notes below.

Solve the following system using the method of elimination:

$$\begin{aligned} -2x+5y&=29 \\ 3x+2y&=4 \end{aligned}$$

Work through Example 4 and take notes below.

Solve the system $\begin{aligned} x-2y&=11 \\ -2x+4y&=8 \end{aligned}$.

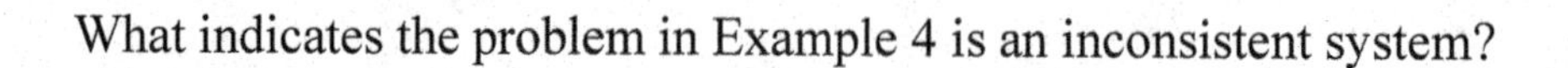

What indicates the problem in Example 4 is an inconsistent system?

Why, geometrically, does the system have no solution?

Work through Example 5 and take notes below.

Solve the system $\begin{aligned} -3x+6y &= 9 \\ x-2y &= -3 \end{aligned}$.

What is the solution of this system geometrically?

What are the two ways the solution can be expressed using ordered pairs?

Section 12.1 Objective 4 Solving Applied Problems Using a System of Linear Equations

Write the Five-Step Strategy for Problem Solving Using Systems of Equations.

Step 1.

Step 2.

Step 3.

Step 4.

Step 5.

Work through Example 7 and take notes below.

During one night at the jazz festival, 2,100 tickets were sold. Adult tickets sold for $12, and child tickets sold for $7. If the receipts totaled $22,100, how many of each type of ticket were sold?

Work through the video with Example 8 and take notes below.

Twin City Foods, Inc., created a 10-lb bean mixture that sells for $5.75 by mixing lima beans and green beans. If lima beans sell for $.70 per pound and green beans sell for $.50 per pound, how many pounds of each bean went into the mixture?

Section 12.2 Guided Notebook

Section 12.2 Solving Systems of Linear Equations in Three Variables

- ☐ Work through TTK #1
- ☐ Work through TTK #2
- ☐ Work through Objective 1
- ☐ Work through Objective 2
- ☐ Work through Objective 3
- ☐ Work through Objective 4
- ☐ Work through Objective 5

Section 12.2 Solving Systems of Linear Equations in Three Variables

Section 12.2 Objective 1 Verifying the Solution of a System of Linear Equations

Define a **Linear Equation in Three Variables.**

Define a **System of Linear Equations in Three Variables.**

If a system of linear equations in three variables has a unique solution, then the solution can be geometrically represented by ____.

Work through the video with Example 1 and take notes below.

Verify that the ordered triple (1, 2, 1) is a solution to the following system of linear equations:

$$\begin{aligned} 2x+3y+4z&=12 \\ x-2y+3z&=0 \\ -x+y-2z&=-1 \end{aligned}$$

Section 12.2 Objective 2 Solving a System of Linear Equations Using the Elimination Method

Systems of linear equations in three variables having at least one solution are called ______________________________ systems.

A system that has exactly one solution is called a ____________________, ____________________ system.

A system having infinitely many solutions is called a ____________________, ____________________ system.

A system having no solutions is called a ____________________, ____________________ system.

Write down the six-step **Guidelines for Solving a System of Linear Equations in Three Variables by Elimination.**

Step 1.

Step 2.

Step 3.

Step 4.

Step 5.

Step 6.

Work through the video with Example 2 taking notes below.

Solve the following system:

$$\begin{aligned} 2x+3y+4z&=12 \\ x-2y+3z&=0 \\ -x+y-2z&=-1 \end{aligned}$$

Work through the video with Example 3 taking notes below.

Solve the following system:

$$\begin{aligned} \frac{1}{2}x + y + \frac{2}{3}z &= 2 \\ \frac{3}{4}x + \frac{5}{2}y - 2z &= -7 \\ x + 4y + 2z &= 4 \end{aligned}$$

Section 12.2 Objective 3 Solving Consistent, Dependent Systems of Linear Equations in Three Variables

If a system has ______________________________ solutions, then the system is consistent and ________________________.

For the two-variable case, a consistent, dependent system can be geometrically represented by a pair of ____.

For the three-variable case, a consistent, dependent system can be geometrically represented by three planes that intersect at a _____________________ or by three equations that all describe the _________________________________.

Work through the video with Example 5 taking notes below.

Solve the following system:

$$\begin{aligned} x+2y+3z &= 10 \\ x+y+z &= 7 \\ 3x+2y+z &= 18 \end{aligned}$$

What is an **identity?**

Section 12.2 Objective 4 Solving Inconsistent Systems of Linear Equations in Three Variables

A system of three linear equations in three variables is inconsistent if all three planes have ____.

Sketch three possible scenarios of three different systems of linear equations in three variables that are inconsistent. In each case, describe the planes.

1.

2.

3.

Work through Example 6 taking notes below.

Solve the following system:

$$\begin{aligned} x - y + 2z &= 5 \\ 3x - 3y + 6z &= 15 \\ -2x + 2y - 4z &= 7 \end{aligned}$$

Section 12.2 Objective 5 Solving Applied Problems Using a System of Linear Equations in Three Variables

Work through the video with Example 7 and take notes below.

While playing a real-time strategy game, Joel created military units to defend his town: warriors, skirmishers, and archers. Warriors require 20 units of food and 50 units of gold. Skirmishers require 25 units of food and 35 units of wood. Archers require 32 units of wood and 32 units of gold. Joel used 506 units of gold, 606 units of wood, and 350 units of food to create his units, how many of each type of military unit did he create?

Section 12.3 Guided Notebook

Section 12.3 Solving Systems of Linear Equations in Three Variables Using Gaussian Elimination and Gauss-Jordan Elimination

- ☐ Work through TTK #1
- ☐ Work through TTK #2
- ☐ Work through TTK #3
- ☐ Work through Objective 1
- ☐ Work through Objective 2
- ☐ Work through Objective 3
- ☐ Work through Objective 4
- ☐ Work through Objective 5
- ☐ Work through Objective 6
- ☐ Work through Objective 7

Section 12.3 Objective 1 Solving a System of Linear Equations Using Gaussian Elimination

What is triangular form?

What is back substitution?

Who was Gaussian elimination named after?

Write down the three **Elementary Row Operations**.

1.

2.

3.

What does R_i describe?

Write down the **Notation Used to Describe Elementary Row Operations**.

Notation **Meaning**

Work through the video with Example 1 taking notes below.

For the following system, use elementary row operations to find an equivalent system in triangular form and then use back substitution to solve the system:

$$
\begin{aligned}
2x+3y+4z&=12\\
x-2y+3z&=0\\
-x+y-2z&=-1
\end{aligned}
$$

Section 12.3 Objective 2 Using an Augmented Matrix to Solve a System of Linear Equations

What is an augmented matrix?

Work through the video with Example 2 and take notes below.

Create an augmented matrix and solve the following linear system using Gaussian elimination by writing an equivalent system in triangular form:

$$\begin{aligned} x+2y-z&=3\\ x-3y-2z&=11\\ -x-2y+2z&=-6 \end{aligned}$$

Work through the video on triangular, row-echelon, and reduced row-echelon form.
Take notes and answer the questions below.

What is triangular form?

What is row-echelon form?

What is reduced row-echelon form?

What is the benefit of reduced row-echelon form?

What is Gauss-Jordan elimination?

Work through the video with Example 3 and take notes below.

Solve the following system using Gauss-Jordan elimination:

$$\begin{aligned} x_1 + x_2 + x_3 &= -1 \\ x_1 + 2x_2 + 4x_3 &= 3 \\ x_1 + 3x_2 + 9x_3 &= 3 \end{aligned}$$

Section 12.3 Objective 3 Solving Consistent, Dependent Systems of Linear Equations in Three Variables

What is a consistent system?

When a system has exactly one solution, then the system is called a

___________________, _______________________ system.

If a system has infinitely many solutions, then the system is called a

___________________, _______________________ system.

Work through the video with Example 4 and take notes below.

Use Gauss-Jordan elimination to solve the system:

$$\begin{aligned} x+2y+3z&=10 \\ x+y+z&=7 \\ 3x+2y+z&=18 \end{aligned}$$

Why is this a dependent system?

How many solutions does this system have?

Section 12.3 Objective 4 Solving Inconsistent Systems of Linear Equations in Three Variables

For a system of linear equations in **Two Variables**, an inconsistent system occurs when the two lines are ______________________________.

For the **Three Variable** case, a system is inconsistent if all three __________ have ____ points in common.

Work through the video with Example 5 and take notes below.

Use Gauss-Jordan elimination to solve the system:

$$\begin{aligned} x - y + 2z &= 4 \\ -x + 3y + z &= -6 \\ x + y + 5z &= 3 \end{aligned}$$

Section 12.3 Objective 5 Determining Whether a System Has No Solution or Infinitely Many Solutions

Work through the video with Example 6 and take notes below.

Each augmented matrix in row reduced form is equivalent to the augmented matrix of a system of linear equations in variables, x, y, and z. Determine whether the system is dependent or inconsistent. If the system is dependent, describe the solution as in Example 4.

a. $\left[\begin{array}{ccc|c} 1 & 0 & -2 & 5 \\ 0 & 1 & 3 & -2 \\ 0 & 0 & 0 & 0 \end{array}\right]$ b. $\left[\begin{array}{ccc|c} 1 & 0 & 0 & -4 \\ 0 & 1 & 0 & 6 \\ 0 & 0 & 0 & 10 \end{array}\right]$ c. $\left[\begin{array}{ccc|c} 1 & 0 & 0 & 3 \\ 0 & 1 & -2 & 4 \\ 0 & 0 & 0 & 0 \end{array}\right]$

Section 12.3 Objective 6 Solving Linear Systems Having Fewer Equations Than Variables

What are the geometric possibilities for a system of three variables but only two equations?

1.

2.

3.

Work through Example 7 taking notes below.

Solve the linear system using Gauss-Jordan elimination:

$$\begin{aligned} x + y + z &= 1 \\ 2x - 2y + 6z &= 10 \end{aligned}$$

Section 12.3 Objective 7 Solving Applied Problems Using a System of Linear Equations Involving Three Variables

How many points determine the equation of a line?

How many non-collinear points determine the equation of a quadratic function?

Work through Example 9 take notes below.

Determine the quadratic function whose graph passes through the three points $(1,-9)$, $(-1,-5)$, and $(-3,7)$.

Section 12.5 Guided Notebook

Section 12.5 Systems of Nonlinear Equations

- ☐ Work through Objective 1
- ☐ Work through Objective 2
- ☐ Work through Objective 3
- ☐ Work through Objective 4

Introduction to Section 12.5

What is a **nonlinear system**?

How can we graphically represent the real solutions to a nonlinear system?

Section 12.5 Objective 1 Determining the Number of Solutions to a System of Nonlinear Equations

Work through the interactive video with Example 1 and take notes below.

For each system of nonlinear equations, sketch the graph of each equation of the system and then determine the number of real solutions to each system. Do not solve the system.

a. $\begin{aligned} x^2 + y^2 &= 25 \\ x - y &= 1 \end{aligned}$

b. $\begin{aligned} x - y^2 &= 4 \\ x - y &= 6 \end{aligned}$

c. $\begin{aligned} x^2 + y^2 &= 9 \\ x^2 - y &= 3 \end{aligned}$

Section 12.5 Objective 2 Solving a System of Nonlinear Equations Using the Substitution Method

What are the five steps for **Solving a System of Nonlinear Equations by the Substitution Method**?

Step 1.

Step 2.

Step 3.

Step 4.

Step 5

Work through the video with Example 2 and take notes below.

Determine the real solutions to the following system using the substitution method.

$$x^2 + y^2 = 25$$
$$x - y = 1$$

After obtaining our proposed solutions to a system of nonlinear equations, what is it absolutely critical to do?

Work through the video with Example 3 and take notes below.

Determine the real solutions to the following system using the substitution method.

$$5x^2 - y^2 = 25$$
$$2x + y = 0$$

Work through the video with Example 4 and take notes below.

Determine the real solutions to the following system using the substitution method.

$$x^2 + 2y^2 = 18$$
$$xy = 4$$

Section 12.5 Objective 3 Solving a System of Nonlinear Equations Using Substitution, Elimination, or Graphing

What are the six steps for **Solving a System of Nonlinear Equations by the Elimination Method**?

Step 1.

Step 2.

Step 3.

Step 4.

Step 5.

Step 6.

Work through the interactive video with Example 5 and take notes below.

Determine the real solutions to the following system.

$$x^2 + y^2 = 9$$
$$x^2 - y = 3$$

Section 12.5 Objective 4 Solving Applied Problems Using a System of Nonlinear Equations

Work through Example 8 taking notes below.

Find two positive numbers such that the sum of the squares of the two numbers is 25 and the difference between the two numbers is 1.

Section 14.1 Guided Notebook

14.1 Introduction to Sequences and Series

- ▯ Work through TTK #1
- ▯ Work through Objective 1
- ▯ Work through Objective 2
- ▯ Work through Objective 3
- ▯ Work through Objective 4
- ▯ Work through Objective 5
- ▯ Work through Objective 6

Section 14.1 Introduction to Sequences and Series

Section 14.1 Objective 1 Writing the Terms of a Sequence

What is the definition of a **finite sequence**?

What is the definition of an **infinite sequence**?

What are the **terms** of the sequence?

What is the definition of the **Factorial of a Non-Negative Integer**?

By definition, we write the zero factorial as _____ and it is equal to ______.

Work through Example 1 showing all work below.
Write the first four terms of each sequence whose nth term is given.

a. $a_n = 2n - 1$

b. $b_n = n^2 - 1$

c. $c_n = \dfrac{3^n}{(n-1)!}$

d. $d_n = (-1)^n 2^{n-1}$

What is an **alternating sequence?**

Section 14.1 Objective 2 Writing the Terms of a Recursive Sequence

Work through the interactive video with Example 2 and show all work below.
Write the first four terms of each of the following recursive sequences.

a. $a_1 = -3,\ a_n = 5a_{n-1} - 1$ for $n \geq 2$

b. $b_1 = 2,\ b_n = \dfrac{(-1)^{n-1} n}{b_{n-1}}$ for $n \geq 2$

Work through Example 3 and show all work below.
The Fibonacci sequence is defined recursively by $a_n = a_{n-1} + a_{n-2}$, where $a_1 = 1$ and $a_2 = 1$.
Write the first eight terms of the Fibonacci sequence.

Section 14.1 Objective 3 Writing the General Term for a Given Sequence

Work through Example 4 and take notes here on the process for determining a pattern in a given sequence.
Write a formula for the nth term of each infinite sequence, then use this formula to find the 8th term of the sequence.

a. $\frac{1}{1}, \frac{1}{2}, \frac{1}{3}, \frac{1}{4}, \frac{1}{5}, \ldots$

b. $-\frac{2}{1}, \frac{4}{2}, -\frac{8}{6}, \frac{16}{24}, -\frac{32}{120}, \ldots$

Section 14.1 Objective 4 Computing Partial Sums of a Series

What is the definition of a **finite series**?

What is the definition of an **infinite series**?

The sum of the first n terms of a series is called the nth _______________ **sum** of the series and is denoted as S_n.

Work through Example 5 and note how you determined the indicated partial sums below. Given the general term of each sequence, find the indicated partial sum.

a. $a_n = \frac{1}{n}$, find S_3.

b. $b_n = (-1)^n 2^{n-1}$, find S_5.

Section 14.1 Objective 5 Determining the Sum of a Finite Series Written in Summation Notation

Watch the video next to Objective 5 and complete the following:

Write down the definition of **Summation Notation:**

Is it necessary for the lower limit of the summation to start at 1?

Fill in the blank:
The infinite series $a_1 + a_2 + \ldots + a_n + a_{n+1} + \cdots$ can be written as ____________.

Fill in the blank:
$a_1 + a_2 + a_3 + a_4 + a_5 + a_6 =$ ______________ .

Work through the interactive video with Example 6 and take notes here:
Find the sum of each finite series.

a. $\sum_{i=1}^{5} i^2$

b. $\sum_{j=2}^{5} \frac{j-1}{j+1}$

c. $\sum_{k=0}^{6} \frac{1}{k!}$

Section 14.1 Objective 6 Writing a Series Using Summation Notation

Work through Example 7 and take notes below.
Rewrite each series using summation notation. Use 1 as the lower limit of summation.

a. $2 + 4 + 6 + 8 + 10 + 12$

b. $1 + 2 + 6 + 24 + 120 + 720 + \dots + 3{,}628{,}800$

Section 14.2 Guided Notebook

14.2 Arithmetic Sequences and Series

- [ ] Work through TTK #1
- [ ] Work through TTK #2
- [ ] Work through TTK #3
- [ ] Work through Objective 1
- [ ] Work through Objective 2
- [ ] Work through Objective 3
- [ ] Work through Objective 4

Section 14.2 Arithmetic Sequences and Series

<u>Section 14.2 Objective 1 Determining If a Sequence is Arithmetic</u>

What is an **Arithmetic Sequence**?

In an arithmetic sequence, the ***d*** refers to ______________________________.

In an arithmetic sequence the **general term**, or ***n*th** term, has the form a_n = _____________.

Work through the interactive video with Example 1 and show all work below.
For each of the following sequences, determine if it is arithmetic. If the sequence is arithmetic, find the common difference.

a. $1, 4, 7, 10, 13, \ldots$

b. $b_n = n^2 - n$

c. $a_n = -2n + 7$

d. $a_1 = 14, \quad a_n = 3 + a_{n-1}$

Refer to the Figure at the end of the solution to Example 1.

Note that every Arithmetic Sequence is a linear function whose domain is the natural numbers.

What does it mean that the points are **collinear**?

When the common difference of an arithmetic sequence is positive, the terms of the sequence__________________ and the graph is represented by a set of ordered pairs that lies along a line with a positive slope.

When the common difference of an arithmetic sequence is negative, the terms of the sequence_______________ and the graph is represented by a set of ordered pairs that lies along a line with a negative slope.

Section 14.2 Objective 2 Finding the General Term or a Specific Term of an Arithmetic Sequence

What is the formula for the **general term of an arithmetic sequence**?

Work through the interactive video with Example 2 and show all work below.
Find the general term of each arithmetic sequence, then find the indicated term of the sequence.

a. 11, 17, 23, 29, 35,…; a_{50}

b. 2, 0, –2, –4, –6,…; a_{90}

c. Find a_{31}.

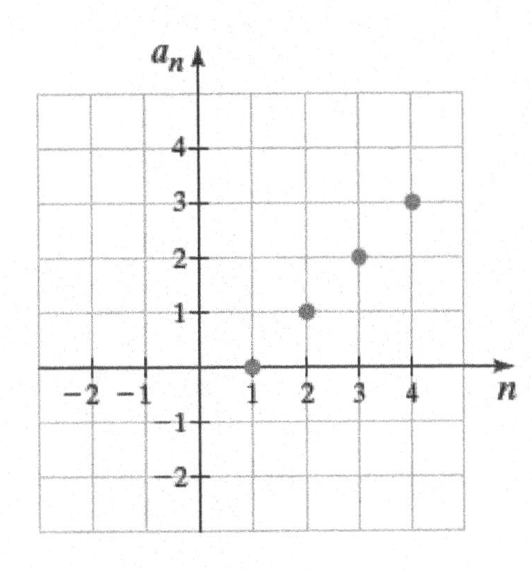

Work through the interactive video with Example 3 and show all work below. Make special note of part b where you are *not* given the common difference.

a. Given an arithmetic sequence with d = –4 and $a_3 = 14$, find a_{50}.

b. Given an arithmetic sequence with $a_4 = 12$ and $a_{15} = -10$, find a_{41}.

Section 14.2 Objective 3 Computing the *n*th Partial Sum of an Arithmetic Sequence

Make note of the formula for the ***nth* Partial Sum of an Arithmetic Sequence**:

Work through Example 4 and show all work below.
Find the sum of each arithmetic series.

a. $\sum_{i=1}^{20}(2i-11)$

b. $-5+(-1)+3+7+\ldots+39$

Section 14.3 Guided Notebook

14.3 Geometric Sequences and Series

- ☐ Work through TTK # 1
- ☐ Work through TTK # 2
- ☐ Work through TTK # 3
- ☐ Work through Objective 1
- ☐ Work through Objective 2
- ☐ Work through Objective 3
- ☐ Work through Objective 4
- ☐ Work through Objective 5
- ☐ Work through Objective 6

Section 14.3 Geometric Sequences and Series

Section 14.3 Objective 1 Writing the Terms of a Geometric Sequence

Write the definition of a **Geometric Sequence:**

How do you determine the **common ratio**?

State the formula for the ***n*th term** or the **general term of a geometric sequence**:

Work through the interactive video with Example 1 and show all work below.

a. Write the first five terms of the geometric sequence having a first term of 2 and a common ratio of 3.

b. Write the first five terms of the geometric sequence such that $a_1 = -4$ and $a_n = -5a_{n-1}$ for $n \geq 2$.

Section 14.3 Objective 2 Determining if a Sequence is Geometric

Work through the interactive video for Example 2 and show all work below.
For each of the following sequences, determine if it is geometric. If the sequence is geometric, find the common ratio.

a. 2, 4, 6, 8, 10, …

b. $\frac{2}{3}, \frac{4}{9}, \frac{8}{27}, \frac{16}{81}, \frac{32}{243}, \ldots$

c. $12, -6, 3, -\frac{3}{2}, \frac{3}{4}, \ldots$

Section 14.3 Objective 3 Finding the General Term or a Specific Term of a Geometric Sequence

Write the formula for the general term of a geometric sequence:
$a_n =$ ________________

Work through Example 3 and show all work below.
Find the general term of each geometric sequence.

a. $12, -6, 3, -\frac{3}{2}, \frac{3}{4}, \ldots$

b. $\frac{2}{3}, \frac{2}{9}, \frac{2}{27}, \frac{2}{81}, \frac{2}{243}, \ldots$

Work through the interactive video with Example 4 and show all work below.

a. Find the 7th term of the geometric sequence whose first term is 2 and whose common ratio is −3.

b. Given a geometric sequence such that $a_6 = 16$ and $a_9 = 2$, find a_{13}.

Section 14.3 Objective 4 Computing the *n*th Partial Sum of a Geometric Series

Note the summation notation for an **infinite geometric series.**

What is the ***n*th partial sum** of a series? Write the partial sum in summation notation.

What is the **Formula for the *n*th Partial Sum of a Geometric Series**?

Work through the interactive video with Example 5 and show all work below.

a. Find the sum of the series $\sum_{i=1}^{15} 5(-2)^{i-1}$.

b. Find the 7th partial sum of the geometric series $8+6+\frac{9}{2}+\frac{27}{8}+\ldots$

Section 14.3 Objective 5 Determining If an Infinite Geometric Series Converges or Diverges

What is the **Formula for the Sum of an Infinite Geometric Series**?

Work through the interactive video with Example 6 and show all work below.
Determine whether each of the following series converges or diverges. If the series converges, find the sum.

a. $\sum_{n=1}^{\infty} \frac{1}{2}\left(\frac{2}{3}\right)^{n-1}$

b. $3-\frac{6}{5}+\frac{12}{25}-\frac{24}{125}+\ldots$

c. $12+18+27+\frac{81}{2}+\frac{243}{4}+\ldots$

Section 14.3 Objective 6 Applications of Geometric Sequences and Series

Work through the interactive video with Example 9 and show all work below.
Every repeating decimal number is a rational number and can therefore be represented by the quotient of two integers. Write each of the following repeating decimal numbers as a quotient of two integers.

a $.\overline{4}$

b. $.2\overline{13}$

Section 14.4 Guided Notebook

Section 14.4 The Binomial Theorem

- ☐ Work through TTK #1
- ☐ Work through Objective 1
- ☐ Work through Objective 2
- ☐ Work through Objective 3
- ☐ Work through Objective 4

Section 14.4 The Binomial Theorem

Section 14.4 Objective 1 Expanding Binomials Raised to a Power Using Pascal's Triangle

Work through the interactive video with Objective 1.
Determine the expansion of $(a+b)^6$ using the Binomial Theorem and Pascal's Triangle.

$n=0$: $(a+b)^0 =$ 1

$n=1$: $(a+b)^1 =$ $1a+1b$

$n=2$: $(a+b)^2 =$ $1a^2+2ab+1b^2$

$n=3$: $(a+b)^3 =$ ______________ (Complete the expansion.)

$n=4$: $(a+b)^4 =$ ______________ (Complete the expansion.)

$n=5$: $(a+b)^5 =$ ______________ (Complete the expansion.)

Note that the sum of the exponents of each term is equal to ______________________.

Fill in the missing blanks for Pascal's Triangle:

$n=0$	1
$n=1$	1 1
$n=2$	1 2 1
$n=3$	1 _ _ 1
$n=4$	1 4 6 4 1
$n=5$	1 _ _ _ _ 1

Explain below how to find the expansion for $(a+b)^6$:

See if you can use the Binomial Theorem and Pascal's Triangle to expand each binomial found in Example 1.

a. $(x+2)^4$

b. $(x-3)^5$

c. $(2x-3y)^3$

NOTE:
The terms of the expansion of the form $(a-b)^n$ will always ____________________ in sign with the sign of the first term being positive.

Section 14.4 Objective 2 Evaluating Binomial Coefficients

What is the **Formula for a Binomial Coefficient**?

Use the formula to work through Example 2.
Evaluate each of the following binomial coefficients.

a. $\binom{5}{3}$

b. $\binom{4}{1}$

c. $\binom{12}{8}$

Section 14.4 Objective 3 Expanding Binomials Raised to a Power Using the Binomial Theorem

What is the **Binomial Theorem**?

Work through the interactive video with Example 3.
Use the Binomial Theorem to expand each binomial.

a. $(x-1)^8$

b. $(\sqrt{x}+y^2)^5$

Section 14.4 Objective 4 Finding a Particular Term or a Particular Coefficient of a Binomial Expansion

The Binomial Theorem can be used to develop a formula to find a particular term within the expansion.

Formula for the $(r+1)^{\text{st}}$ Term of a Binomial Expansion

If n is a positive integer and if $r \geq 0$, then the $(r+1)^{\text{st}}$ term of the expansion of $(a+b)^n$ is given by

$$\binom{n}{r} a^{n-r} b^r = \frac{n!}{r! \cdot (n-r)!} a^{n-r} b^r .$$

Work through the video with Example 4 and show all work below.
Find the third term of the expansion of $(2x-3)^{10}$.

Work through the video with Example 5 and show all work below.
Find the coefficient of x^7 in the expansion of $(x+4)^{11}$.